D0773538

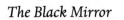

The Black Mirror

The Black Mirror

Looking at Life through Death

RAYMOND TALLIS

Yale UNIVERSITY PRESS

NEW HAVEN & LONDON

First published 2015 in the United States by Yale University Press
and in Great Britain, as *The Black Mirror: Fragments of an Obituary
for Life,* by Atlantic Books, an imprint of Atlantic Books, Ltd.

Yale University Press books may be purchased in quantity for
educational, business, or promotional use. For information,
please e-mail sales.press@yale.edu (U.S. office) or sales@yaleup
.co.uk (U.K. office).

Printed in the United States of America.

Library of Congress Control Number: 2015940159
ISBN 978-0-300-21700-1 (cloth: alk. paper)

A catalogue record for this book is available from the British
Library.

This paper meets the requirements of ANSI/NISO Z39.48-1992
(Permanence of Paper).

10 9 8 7 6 5 4 3 2 1

To Julian Spalding – in friendship, gratitude and admiration

Acknowledgements

It is a pleasure to acknowledge the enthusiastic support of James Nightingale and Margaret Stead at Atlantic for this book, the scrupulous editorial work of Luke Brown and the guidance and inspiration of Toby Mundy over many years.

Contents

Lucem demonstrat umbra
The darkness shows forth the light

Mutato nomine et de te fabula narratur
Change only the name and this story is about you

—HORACE

Overture: In the Beginning Was the Word

Death destroys a man; the idea of death saves him.

E. M. FORSTER[1]

Let us banish the strangeness of death: let us practise
it, accustom ourselves to it, never having anything so
often present in our minds than death: let us always
keep the image of death…in full view.

MICHEL DE MONTAIGNE[2]

Death is nothing – a limit to the life it is not. Those whom we call
'the dead' neither enjoy their peace nor endure their loss. But if death
were merely nothing, there would be nothing to be said about it and
this book would comprise as many blank pages as the reader could
tolerate. But death is less than nothing. It is an omni-ravenous zero.
Like the God of certain theologies, death is defined by what it is not.
Unlike that 'apophatic' God, however, it has a parasitic being, borrow-
ing from what it subtracts or destroys, acquiring apparent substance
by feasting on life. This book, therefore, is about what death takes
away. It is an obituary, or fragments of an obituary, for life.

While death preys impersonally on the entirety of the living
world, it is also deeply personal. Each journey from I-hood to

thing-hood and onwards to loss even of thing-hood is unique. This transition from somewhere to nowhere, from someone to no-thing, is beyond the grasp of particular thoughts entertained by a particular person on a particular morning, afternoon, evening or night. Truly to think your extinction, you would have to become the equal in your thoughts of the sum total of yourself that is cancelled. So, while fear may be important, something deeper than fear stops us fixing our attention on our end. It is its inconceivability.

Fluency, that by dint of onward momentum precludes descent into the depths, is a constant temptation. 'Death' after all is a word we use lightly: 'I cut him dead', or (of an embarrassment) 'I died a thousand deaths' or (of a hangover) 'I felt like death'. And there is no shortage of remarks attached to Impressive Names, apophthegms, *obiter dicta*, jokes (gallows humour for those for whom 'gallows' is just a word) or consoling, or monitory, reflections, ready-made phrases gesturing towards feelings that are actually unfelt and thoughts that are unthought.

Indeed, talking about death may be even more evasive than remaining silent. It is not difficult to assume a portentous voice that, drawing on the noise this side of the grave, purports to speak for the silence on the other. Or to seem to outsmart death by putting it in its place, usually by personifying it to rob it of its 'sting', to deny it its 'victory'. Tell that to the – dead – Marines.

Actual dying and real bereavement expose the hollowness of these triumphs. Few aphorisms – and a few keystrokes are sufficient to liberate a torrent – sound good when uttered between retches, bracketed between groans or spoken in a world emptied by the loss of one's life's companion. Anything that does not pass W. H. Auden's test – 'something a man of honour, awaiting death from cancer or a firing squad could read without contempt'³ – is mere bling. But this is an impossibly high bar.

If death cannot be reached, even less tamed, by speech – and to attempt to think or speak of it sometimes feels like trying to breathe in a vacuum – what purpose could be served by a book devoted to it? Have we not, after all, been misled by the existence of a word into thinking that there is an item, a subject, a topic, to be discussed?

What is to be discussed is life. *Lucem demonstrat umbra* – 'The shadow reveals the light' – says it all. The unspeakable Nothing italicizes at least some of the Everything that is life. While death destroys us in fact, the thought of our non-existence may save us from triviality, from entrapment in secondary things – only temporally of course, but then life itself is a temporary matter. To be oblivious of death is to be only half-awake.

This is an implicit rejoinder to Spinoza's assertion that 'The free man thinks about nothing less often than about death, and his wisdom is the preparation not for death but for life.'⁴ The free man (and woman) who is preparing for life may think more deeply and, indeed, more freely by thinking about death. In order to live like a philosopher, it is necessary to die like one – that is to die in thought and in imagination before you die in body. Few, if any, can philosophize while panting for breath, or vomiting, and none while confused, or comatose. No argument or revelation will save me when, as will surely happen, I shall be utterly broken and my body will embark on a one-way journey to extinction. No sentence will reach to the bottom of my grief, my pain, or my nausea. And this is why Montaigne enjoined us to 'banish the strangeness of death' and 'always keep the image of death in our minds and in our imagination'.

The attempt to see life from the perspective of death brings us close to the heart of the philosophical impulse. 'Philosophy', David Pears said, 'originates in the desire to transcend a world of human thought and experience, in order to find some point of vantage from which it can be seen as a whole.'[5] Death, or the idea of it, may seem to offer that vantage point – not as an Archimedean fulcrum, but as an imaginary outside which makes life (with all its outsides and insides) seem to be an interior, and ourselves to be exiles pressing our faces to a window, watching what is happening within. If to be a philosopher is to be an onlooker, the vantage point of death is the *ultra ne plus* of the philosophical viewpoint: you look upon your life from the virtual position of one who has outlived it. Death's obsidian surface is a rear-view mirror reflecting the life that is now over. We stare at the darkness in order to see more clearly the life that we think of as behind us.

Such a perspective, liberated from the parochial interests of everyday life, is true to the strangeness of our being in the world. We forget this strangeness when there is so much living do to and living demands of us a responsible, focussed attention. How can I be astonished at the miracle of a Wednesday afternoon, of a city, of a conversation, when there is shopping to be unloaded from the car, a toddler walking towards a plug-hole with extended fingers, a clinic to be completed, and a thousand other things justly expecting our considered response?

And so *The Black Mirror*, which reaches out to death but cannot reach it, is about life. Hence the sub-title: *Looking at Life through Death*. Its ambition is to render one's self, daylight, parts of the world, *contre jour,* to revivify the life we have lost in living, to express a deep gladness at being alive, mindful that

In the cold waters of Lethe you will remember that the warm earth meant a thousand heavens.[6]

If death disappears from sight for long stretches of this book, therefore, this is not through oversight. Death is the viewpoint and the viewpoint is not part of the view, especially when that viewpoint is eyeless beyond even that of the mathematical representation of the world.

If I had had more courage, imagination and power of concentration, this book would be more obviously what it is: a walk across a tightrope that has nothing to hook on to at the other side because there is no other side.

One does what one can.

References

1. E. M. Forster, *Howards End*.
2. Michel de Montaigne, *Essais*, 1580, M. A. Screech (ed.) 1991m, Book 1, Section 29, p. 96.
3. W. H. Auden, 'In the Cave of Making' (In Memoriam Louis MacNeice).
4. Spinoza, *Ethics*, 4, Prop 67.
5. David Pears, *Wittgenstein*, Fontana Modern Masters (London: Fontana Collins, 1971) p. 21.
6. Reiner Schurmann, paraphrasing Osip Mandelstam, 'Brothers. Let Us Glorify Freedom's Twilight', quoted in Vishwa Adlur's *Parmenides, Plato and Mortal Philosophy: Return from Transcedence* (London: Continuum, 2011).

PART ONE | Ending

one

To the Sunless Land

While of death, there is little to be said that is truly to the point, of dying there is too much. Nevertheless, if death is to help us to see the life that it subtracts, the unavoidable first step must be to look at the process of subtraction; at the journey to The Sunless Land.

Though it seems strange that something as undifferentiated as death should have routes or names, there are countless paths from the person to the corpse, so many *fleurs du mal* strewn in its way, so many ways of being put on notice that the end is coming from a definite direction, so many ways of experiencing and enduring the journey from the 'I' or 'me' or 'you' or 'he' to an 'it', from someone to something, which continues on to another road – to nothing in particular, nowhere in particular.

Different organs, or different afflictions of different organs, may take the lead; the processes may be visible or invisible, audible or silent, odourless or stinking. While death is certain, we rarely know in advance which of the thousands of pages of the medical text-books will describe the portal through which we will be expelled from the world. For we inhabit only the coastline of our body (and even this only patchily – when did you last have anything to do with the creases on the back of your neck or the down on the small of your back?). The object that goes under our name is, for the most part, as opaque to us as the rest of the material world. We note the reactions of our bodies as if we were external observers. We wonder at the tingling of our hands, the rumbling of our guts, and the twitching of our calves. We await the dripping of sweat as a message that we have done sufficient of the exercise that we hope might postpone our end.

The signals from the hinterland may be difficult to interpret. One day, as RT walked up a hill and felt a pain in his chest, a question formed in his mind: 'Indigestion or angina?' He could not differentiate between a minor discomfort and a harbinger of the possible end of his world. And once, when he was sitting on the toilet, reading the *Times Literary Supplement,* waiting to be distanced from the contents of his colon, he noted a dark spot on his thigh, and speculated idly as to its meaning. Would a sky-cancelling wing unfurl from this macule of darkness?

Death is a potent reminder that we do not live, fill, and animate the facts of our case. They are bigger than we are and smaller than our experiences. And the mode of our death often has little to do with the biographical facts, with the curriculum vitae: in dying we have to live out impersonal processes that are necessary for, but are far beneath, the lowest stratum of our person. The opaque philosopher Jacques Derrida, the parliamentary sketch writer and delicious wit Simon Hoggart, the brave, visionary GP Anne McPherson, and the wife of a friend of a friend of RT had little else in common except that they were all fatally assaulted by a pancreas which broke out of its bounds and ransacked their bodies. Tennessee Williams' death from choking on a cap from a bottle of eye drops seems to contain little of the Deep South – or his journey from the Deep South to international fame via Broadway.

The disconnection between our death and our life is cruelly underlined by those who die young, whose life story is broken off after a few preparatory sentences. 'It doesn't make sense,' we say. But it makes precious little sense at any age; that, for example, a man of 90, taking his usual stroll and looking forward to a pint of beer, should be felled by a clot in his lungs, and consequently be translated at once from a particular time, place, world, and life, to no time, no place, no world, and no life. Death is not a neat full stop at the end of the final sentence, of the final paragraph, of the final chapter, of a life. It is the profoundest of all interruptions.

Dying takes you deeper into the inscrutable, lampless hinterland of carnal being. More of your body becomes vocal, gatecrashing what it feels like to be you, though the 'you' is transformed to something more general than the you of your loves and hates, joys and sorrows, of your virtues and vices, your hobbies and duties, your CV. Dying is a mixture of subtraction – the breaking of links, the narrowing of scope – and of addition – more effort, malaise, pain, and nausea, more general noise attending daily life. Gilded memories or a roseate future are pushed to one side by an ever more obtrusive soiled present until the dying man experiences himself as a gobbet of carnal rubble in the waste bin of his world.

How death will come is uncertain; that it will come is certain. The necessity of our death follows from the nature of our life. We die because we are improbable. Something as highly structured as our body is at odds with the overall tendency of the universe. The beautiful order of our faces, our hands, our hearts, is won from a sea of disorder whose overall trajectory is towards dissipation.

The emergence of life remains, for this reason, a mystery. The increased complexity of what we call 'higher' life forms is an additional mystery. Competition, selective pressures acting on mutations, only seem *post hoc* to explain the journey from prebiotic crystals, lacking even membranes, to men in suits and women in dresses, from mute RNA to individuals like RT who cultivated a sense of life's little ironies. Even if consciousness and self-consciousness gave living creatures an advantage over the dimmer competition (and it's not at all clear that they do) this would not explain why the material world should acquire consciousness, be conscious in parts of itself, as itself, self-aware to the point where it says 'I'. (None of this, by the way, amounts to a case for an 'intelligent designer'. The gaps in our understanding don't add up either to the argument for, the job description of, or the shape of, something corresponding to the word 'God'. God is not only laden with historical baggage: it *is* historical baggage.)

We are vulnerable because we are complex organisms – though some of our complexity is devoted to mitigating our vulnerability – and the physical world abhors such complexity. Less specifically, we are condemned to transience because we are the children of change. A restless universe gave rise to us and that restlessness is governed by law. It makes no exceptions. We cannot expect our birth to open up a parish in the material world in which the universe ceases to be restless and ignores its own laws. That by which we are swept into existence is that by which we are swept out of existence.

And, just to make sure there is no escape from extinction, we rely on the restlessness and laws to be maintained during our (long, short) lives in order that there shall be a finite, rather than infinitesimal, interval between the first cry and the last; the gap between the entrance and the exit called 'my life' should be a matter of years rather than nanoseconds. If the ordinary processes of the world were suspended for our convenience, then food and water and shelter would have no power to protract our being. The effectiveness of basic means to life belong to the Great Mechanism that is the universe, that unmakes us by the same means by which it makes us.

Ultimately helpless, we nonetheless take a hand in managing our mortality. Much of the business of our life, when we are not taken up with pleasures and diversions, serves the overriding project of postponing death, of fending off the accidents of nature, to which, being accidents of nature, we are prone, and with making the world a more hospitable place and our bodies more aligned to our personal narratives, to the lives we have chosen to live rather than merely the processes that make those lives possible.

The variety of postponements is astounding. Of course, we eat and drink, and shelter ourselves in clothes and dwellings. Underpinning these basic activities there is a massive infrastructure of practices, skills, technologies, customs and laws. And postponement of death takes more indirect forms than securing nutrition, hydration, and protection. Man is the precaution-taking animal. Hoarding and storing, barricading and padlocking, pacifying the natural world and regulating the human one, are just some of the many ways in which we try to make our lives nice, humane, and long as opposed to nasty, brutish, and short. We look to deflect not only the enemy without – wild animals, infestations, cold, heat, floods, storms, volcanoes, and, most terrifyingly, our fellow men – but also the enemy within. Immunization, a balanced diet, exercise, moderation in pleasures, and pills and operations, are some of the conduits through which a vast amount of knowledge is mobilized, often via dizzyingly complex modes of cooperation, to arrest or postpone the various processes (visible and invisible) that make up the passage from ourselves to no one.

The appointment with extinction is postponed but not cancelled because the very process of living is inseparable from death. Most obviously, use wears out parts, toxins accumulate, and the mechanisms of growth may produce growths. Accidents that make us happen will make us unhappen: we are composed of elements that otherwise have little to do with our lives. Life grows out of – and feeds on – death, though it is sometimes difficult to get life to believe this.

Indeed, death's outriders are ignored or resented as mere interruptions. Its every manifestation is at an impertinent angle to the ongoing stories of RT's life; an anti-project getting in the way of his projects. The appointment with the doctor, the attack of vomiting, even the summons to the funeral of a friend, clash with the many businesses of life. But the invasions of irrelevance persist, the drumbeats get louder. Sooner or later, death becomes the main, overwhelming business and dying the only story. RT will turn to the left and It will be there with folded arms. He will turn to the right and It will be waiting with an implacable smile. All escape routes will be locked, barred and bolted and the keys – his dying body – will melt away.

The journey from the person to the corpse, from 'I' to 'it' may be a plummet or a long, winding, downward path, a howling descent through thorns, an exit from the world through an endless razor-wired fence.

Sudden death – however much we may wish for it as preferable to being racked by our own body – seems more shocking. It is a reminder that we are never at any time insulated by a guaranteed distance from the end of ourselves. Death has no obligation to serve notice. An ordinary Wednesday with no cosmological, apocalyptic or even philosophical pretensions may be the date when RT's world is extinguished. The afternoon in which his life reached its total looked pretty much like its predecessor the week before whose main event was that he went shopping with Mrs RT and agreed, with slightly more warmth than the occasion called for, to take a taxi. Sudden death is doubly shocking because it allows no opportunity for farewells, for settling one's affairs, for tidying up behind one, for saying that which has to be said, and concealing that which has to remain hidden.

And it can pounce from so many directions. You are abroad. Forgetting that you are not in England, and preoccupied by something you are planning to say at the meeting you are due to address, you walk in front of a car, driven on what for you is the wrong side of the road, by someone who was also a little preoccupied. Or you are caught in the crossfire between two individuals bent on killing each other. Or you come into your spouse's study and say 'I have a terrible headache' and those are your last words. On impulse you go for a late swim and are swept out to sea. The thousand conversations are broken off, never to be resumed. The photograph for which you impatiently posed turns out to be the last that was taken, and the smile an incomplete triumph over irritation at being told to smile, the last to be recorded.

The simplicity of sudden death mocks the exquisite, painfully constructed complexity of the life that it ends. Surely, we feel, it should take time to unpick all that was so carefully woven together: all that one's parents, teachers, mentors, civil society, technology, science, and life itself put into enabling a world to be constructed which could navigate the Great Outside in which we find ourselves when we are pitched into our lives. The mismatch between the difficulty with which we are put together – the love, patience, and painstaking concern necessary for our flourishing – and the ease with which we can be torn into meaningless pieces is shocking. Think of all the many, minutely detailed anxieties RT's parents had for him – Is he eating enough? What does that fever mean? Is he safe on that swing? Is he being bullied? Will he pass his exam? Will he get the job he wants? Will he be happy with Mrs RT? Is he going to be promoted? Can we help him to worry less about his child? All the care, nurturing, vigilance, protection, education leads like a long upward slope to a cliff face. You have learned to 'Keep away from the edge!' but sooner or later, the edge, built into the very stuff of your lives, comes for you. A bang on the head and RT falls through all the storeys and stories of his life to a condition less than that of the lowest of the beasts, one that lacks even the order granted to a crystal.

There are modes of dying that, however distressing, seem at least to do justice to our complexity by removing those storeys and stories one by one. Dementia – Jonathan Swift's 'dying from the head down' – unscrambles all that we have received or have achieved in reaching our state as fully developed human persons. The learning curve points remorselessly downwards and we forget all that we knew, understood, and could perform: knowing-how goes the way of knowing-that. Occasional tatters of clearness reveal by default the personal universe dissolving in the thickening, ubiquitous inner fog. But mostly we are puzzled, lost, in an anguish of alienation. The person with whom we have spent decades in multilayered intimacy becomes a stranger inexplicably in our lives, looming, and lurking, as we grow ever stranger to them, though for them this estrangement does not puzzle because it has a general name – Dementia – and follows a general pattern: the tragic picking apart of all connections. The sufferer is the one least able to grasp that the changed and frightening and bewildering and frustrating and lonely world is an unchanged reality refracted through changes in himself. The word 'Alzheimer's' that he may have used in the past means as little to him as do his once-beloved children. They are now nameless, without recognizable faces, and shorn of their lives – the lives he fostered and shared and the lives they have developed independently of him. He is the one least able to understand the link between the loss of the connectedness between his brain cells and his currently being called 'Poppet' (rather than 'Professor') by a carer, paid to care, whose fingers he is trying to break as she changes him out of his soiled pants. As RT is progressively leached out of the body that bears his name, it becomes a mobile gravestone marking his absence.

The demolition of the mind – and subsequently of the body – brick by brick, is hardly an attractive alternative to sudden death. And it is true that the description of any slow death 'after a long illness' (often 'bravely borne') is usually a euphemism. The battle against the processes that gradually picked him apart was not a fight chosen by the combatant; rather it was an intimate civil war within his body which he suffered more than he engaged. There is a pressure to go through Hell for others: 'We are going to beat this together.'

The phrases – 'long illness', 'bravely borne' – allow us to look past the fear of pain, disability and disintegration, past the disappointed hopes, past the gradual withdrawal into the solitude of an ever sicker, limited, body. They look past the return of the constraints of toddlerhood, this time without a future to look forward to, past the days marked by the helplessness of childhood waiting while fitting into others' busy lives ('They have their own lives to live'); past the continuous dependency on others' good will, and the burden of being a burden, however willingly shouldered; past the stripping off of all dignity. The tears of frustration, the incontinence and the impotent rages. These brave phrases allow us to forget the months doled out in painful seconds; the room, the chair, the bed, as a craft rising and falling on the swells of nausea; the walls closing in as every action becomes ever more difficult; the dreary routines of pressure-sore cleaning and re-siting drips, or eating food made inedible by a sore mouth and rebellious gut; the fatigue invading limbs, trunk, and head, such that every action is a race up a sand dune, until defeat is conceded; the slow closing down of the world to walls identical with the limits of the stricken body. These brave phrases gloss over the way breathing becomes an activity in its own right, not the presupposed, disregarded, background of everything he did. They look past the weary hours in the night of wakefulness alone in the darkness, cowering in anticipation of the lightning strikes of bones eating themselves, of pains that he cannot withdraw from, of shivering and sweating, or of chasing shadows that are transformed into evil spirits last encountered in the bedtime terrors of childhood. They overlook the pruning of possibility and shrinking of the world: the last holiday abroad, the last journey to London, the last shopping expedition, the last hour in the garden; the last time the stairs proved climbable, the last independent journey to the bathroom, or he could sit out in a chair or sit up in bed without assistance. Country-fast, city-fast, under house arrest, room-bound, chair-bound, bed-fast. Thus the landmarks of human being turned from becoming to begoing.

There are other, gentler routes, the road of frailty, of gradual, even painless attenuation. The downhill path of the 'dwindles' that single out no particular organ. The imperceptible diminution of strength such that he no longer simply sat down but lowered himself, or docked, into a chair and no longer leapt or even merely got up but raised himself by main force to a standing position in which he remained stooped towards the earth awaiting him. Dying of old age that does not hurt or feel sick or scream with terror. Doddery, tottery, yes. A little short-winded. Daily activities slightly more difficult. An increasing vagueness of mind and diffuse incapacity of body; a de-differentiation that anticipates the return to the journey, shared with the universe, towards thermodynamic equilibrium. And then, one day, a heart attack that passes almost without notice because the pain fibres serving the heart have themselves aged and have become a little less attentive.

Thus the journey from that spot on the skin, that pain, that lump, that cough, that mysterious weight loss, into a world whose horizons are drawn ever tighter, to the final collapse of the space that had been opened up by his awakening senses, to the point where he could no longer see or feel his hand in front of his face or see or feel that he could not see or feel them. RT goes down with the organism to which his life is fastened and which, failing, changes from a craft which enabled him to voyage across the world to the sea water in which he drowns; changes to a darkness in which the captain, the ship, and the ocean are all one. To the end of an ordeal that will not be over until he, too, is over, and consequently unable to feel the relief that his end will have brought.

Death cannot be avoided so RT hopes for an ideal death: out like a 100-watt bulb at the age of 100. Or fading like a summer evening with birdsong gathered quietly into nests and birds gathered into sleep, the darkness swallowing them all in safety. An end whose last words are: 'Goodbye my darling. How lucky we are to have lived together, to have loved each other. And thank you, world, for having me' – as the dark as well as the light is extinguished.

A death whose tragedy is blunted. But it is a tragedy nonetheless, for which we are existentially unprepared. We have been present since the beginning of time because we have never known – except by report – a time when we did not exist, never lived in a world uninhabited by ourselves. We cannot imagine everything being taken away from us, including the self who enjoyed, suffered, loved, loathed everything that has been taken away. We cannot conceive of the absence of ourselves.

Yes, we know objectively that the reasons we came into being are not sufficient to sustain creatures like us forever, even less the particular lives we lived and shared with those we loved. But we cannot think that we are truly finite and that we shall, after a life of incomplete meanings, cease to be. We hope that, at the very least, our lives will come to an appropriate end when all that we hoped to achieve will have been achieved, that the story will be rounded off, that all passion for existence will have been spent. That we will not be sent on our journey to dust, and to something less organized than dust, until the stories we started in our lives will have come to a satisfactory conclusion.

This will not happen. There are too many stories, too disorganized, that have too many characters, interfering with each other, for any lived story (other than those that are committed to words) to

retain its identity even through a charmed (that is to say uninter-
rupted) journey from beginning to end. And there are new stories
opening up all the time. In many cases, these are initiated by ourselves as we find new projects, new passions, new discoveries about
ourselves and the world. The most cheering thought, however, is
that the world lacks narrative neatness because it is too rich with
abundance pouring in through the sides of any narrative.

And so all that remains to us is to step back and see the world that
has been lost in its inexpressible, untidy richness. But first we need to
reach the zero position from which the world can begin to be seen
and life's obituary to be written. To see what remains after RT has
died. RT, therefore, must pay his last respects and circle round an
item that, innocent of its discourtesy, will not reciprocate his gaze.

His corpse.

two

Last Respects

Disrobe, RT, and look at yourself in the mirror. The image is of an earlier time-slice of the item that will be your corpse, a time slice that, unlike the later one, is looking back at you. Pinch the back of your left hand with the thumb and index finger of your right hand. What in future will be your corpse is pinching itself: the same fingers, the same arms. The face that facing yours is the one that will be exhibited to those who have come to pay their last respects. You are exchanging glances with the past tense of liquefying carrion or a handful of ashes.

Touch your neck and feel the bumping of the carotid artery beneath your finger, at once alien and deeply familiar, marking your personal time in an impersonal universe, signifying the living machine getting on with its business so that you can get on with yours. This is where, when the time comes, nurse or doctor will establish that you are pulseless and no longer an issue for yourself. Sit down and feel the pressure reporting the action and reaction of your buttocks and the seat: this is the weight of the body they will lift on to the trolley headed for the mortuary, a weight that will no longer be 'my weight'.

The present 'You' and the future 'It' share so many predicates: the colour of the eyes, the sharpness of the nose, the shiny cranium, the scar in the sole of the foot. The flesh you are touching will remain you – the you who is presently looking at yourself, hearing the birdsong in the garden, the traffic passing in the road outside – so long as it is sufficiently close to parameters of physiological normality, parameters of which you have little inkling. After this, the 'it' that you are and are not will outlast you for a short while, and as 'the body of the deceased' be invested with the after-image of the extinguished you, holding down a place in the material world for others to view. The body of the deceased will be yours – 'my' corpse – only in prospect, since in reality it will be no one's, setting aside the legal fiction that it belongs to the State to which in a sense you belonged as a citizen while you were alive.

You come to it – this prospective, imaginary viewing of your corpse – from different places in your life. From an early morning in winter, a desk lit by an Anglepoise lamp; a summer's thought in a meadow between glimpses of high clouds propelled by inaudible breezes processing from horizon to horizon, encircling wide skies; between patients on a seasonless ward round, when you are preoccupied by words, charts, arrangements, responsibilities; looking past the drowsy head of a little child as the bedtime story comes to an end. You have tried to glimpse it from your childhood, boyhood, youth, maturity, parenthood, middle age, late seniority, where you momentarily thought of your end but could not sustain the thought because you were distracted by the multicoloured world that said 'Look at me' or because you had more pressing things to do.

Frustrated by your own inattention, you are tempted to sting yourself into sharper awareness by spelling out brutal truths that verge on insults, describing this body when it is no longer anyone's body as 'a carcass', 'meat', 'carrion'. Anything to remind the 'he' that is now of the 'it' that it will be. But still you visit your unselved future body only fleetingly. And you relish walking away, now as free as those who in your last illness, your final days and hours, will have come to see you, trailing fragments of the world they have temporarily set aside – the weather in the streets written on their cheeks, a briefcase stuffed with the day's work, gossip, and a present (a novel that you will not finish); or, more dramatically, will have flown in from abroad, summoned by news of your decline. You imagine those legates from the great world converging on a small room, a hospital bed, staying briefly, and then diverging again into those wide open spaces from which, shrunk to your suffering body and the thin penumbra of care and concern and to-ings and fro-ings that surround it, you have involuntarily withdrawn.

The corpse you will become will be beyond all that. The tubes will have been withdrawn, the last futile tablets taken, the last reluctant mouthful of food, the last sip. Lying in stasis. While you are longer and more corpulent than you were the day you came into the world, life-soiled as opposed to new-minted, you are nonetheless as naked and as lacking in estate. Homeless, propertyless, wifeless, childless, friendless, jobless, thoughtless, breathless, pulseless, gazeless, and so completely sensationless as not be able to experience even numb-ness. Still.

Resting in peace but neither restful or at peace. Behind those closed lids there are no thoughts, dreams, or sleep. Soundlessly stiffening beyond the possibility of movement (a rigor reflected in the creases that can't be smoothed out), inert beyond passivity, its immobility has none of the luxury of repose you once enjoyed.

Relaxation brings none of the deliciousness of relief from labour, from effort; it is no more at rest than a pebble is at rest. Its silence is beyond any lived silences; it is a solid, dense, unheard and ownerless silence; the thoughtless silence of the material world, out of shot of all ears; beyond the seemingly interminable chuntering, muttering, pondering, mumbling, puzzling, and rehearsing of the voice-over, which had footnoted his every waking hour; beyond the faintest susurrus of his quondam self-presence. The taciturnity of the dead is absolute; the analphabetic silence beyond even the Zs of sleep. An amnesia of perfect completeness that has forgotten that it was once RT; the craft has dissolved into the ocean on which it journeyed for so long.

All parts of his body – brain, eyes, and toenails – are now meta-physically equal. They have fallen together to the bottom of the ontological heap that places humans at the top, rocks at the bottom, and bricks, ants and trees inbetween, items that have in common an extensity that physics explores with a radically democratic gaze, allocating all things to a single mode of being, variously described by push-pull mechanics or by atoms that evaporate on closer inspection to mere numbers. The parts that were invisible to him, the below-stairs staff serving organic functions – hollow and solid viscera, pumps and bellows, bones and joints – are on a par with those that formed RT's appearance, by which he was recognized – his face, and hands, and skin. All are carrion: the cheek with the crease across it that will never be smoothed out and the coagulated eyes behind the lids as much as the spleen or the intestine later being sluiced by a hose in the post-mortem room, where some of his mortal parts, donated to science, may be inspected by science.

★

Deserted by agency, RT suffers allcomers to do unto him what they feel is right. His arm, notwithstanding that it casts a shadow similar to the one it cast in life, does not rise unless they lift it; neither do his eyelids close without their gentle touch. And so they attend to him: the laying out, punctuated by the occasional asymmetrical kiss of warm sentient lips on a cold cheek insentient as stone, or a gaze as unreciprocated as that from a statue. He is not there to receive their tears. He is not there.

Before the formal viewing, there is the laying out, the last offices – 'to ensure that the care you have given him in life is carried over into death'. Items that had not been an integral part of his body, though they had been assimilated into his being in the world, are removed and put into labelled bags. The spectacles that had sharpened the world he looked at have already been taken off. There is no hearing aid, though his final years might have been less irritating to others had he taken the trouble to make it easier for him to hear them. The teeth are his own. And his wedding ring, slipped off by the nurses, shares a polythene bag with his spectacles, locked up in the ward safe, before being transferred along with the last pair of pyjamas and that unread novel to the one who is now its owner.

And then the laying out proper – washing, oral hygiene, brushing his beard (his hair having long since deserted his cranium). This passivity is not entirely new. The last few weeks of helpless dependency had brought to an end the years in which he had washed himself and, as both agent and patient, actor and substrate, had scrupulously absterged the encroachments of the environment and removed the adherent by-products of his body's self-maintenance, the secretions and excreta necessary to ensure the dynamic equilibrium which had served him for so long. The hands, foxed with spots, that had washed each other, as well as the body of which they were the chief agent, had been washed by other hands, as he and they lost their grip.

An anal plug has been inserted to pre-empt posthumous bowel actions, though he would have felt none of the shame with which he had once been familiar when surprised by the contents of his own body. Being cleaned is no longer associated with a sickly sense of regression to babyhood. Unclothed, he does not feel his nakedness as exposure. Beyond self-disgust, he is not repelled by the putrid thing his body had started to become.

Even so, for the present, his body is treated with the reverence due to the person he had once been. His head is raised up with care, so that the pillows can be re-plumped. His arms are lifted and lowered gently so that they can be folded across his chest and his hands clasped together, with supportive pillow at the elbows, reinforcing the image of a marmoreal stillness effacing the memory of the wildness and horror of the final hours and days. The chin is propped up to prevent the mouth falling open in a silent scream and the lips are coated with Vaseline to keep them moist, the nurse's finger a proxy for the tongue no longer able to lick them. The eyes are closed with care – to conceal the emptiness of the gaze – without damaging the delicate tissues of those lids that had served up little doses of visual oblivion and wiped cleansing tears over weary hours.

Finally, the room is prepared for visitors. The sheet is smoothed out and the space around the bed bears no record of the hi-tech clutter that had surrounded him for so long. On the bedside table a vase of flowers, silently sipping water, withers next to him. He has said farewell to himself – though it was not 'Goodbye, RT' or a last kiss – but a gasped-for breath; now it is time for others to say farewell to him.

Those who enter see his head protruding from the snow-barrow of the sheets. His head had been it*self* in a way that that goes beyond the impoverished identity relation every item has with itself, as when a brick is a brick, though this, the capital of his body, no longer knows it. The head had been colonized by RT's awareness to varying degrees at different times under different circumstances. It looked at, listened to, smelt, tasted, and touched itself. The gaze looked at the face and judged it as satisfactory or unsatisfactory. The ears enjoyed hearing the mouth whistling in a certain way. The nose inhaled the scent of sweat, secreted in response to exercise undertaken in order to postpone the present state, to dodge the invisible bullets, the emissaries of death. The tongue tasted the spilled blood from the accidentally bitten flesh. The hand felt itself, its opposite number, the beard it stroked, the small of the back where the itch was just out of reach. The toes had had a dimly sideways awareness of each other mediated by haptic glances, accentuated in tight shoes. A maculate pattern of warmths and coolths, a rash of cutaneous sensations, were co-present with aches and joys from below decks, rumours from the alien darkness within, that was and was not his darkness, and not a murmur from his brain.

Between these localized bodily self-revelations had been a continuum of low-level carnal consciousness into which they were located – something more intimate than a body image, more filled in than a mere body schema. The tingle in the toe, the chair pushing back on the buttocks reporting the weight of the torso, the buzzing in the ears, the vanishing point behind the eyes, had all been natives of the same country, united and yet scattered, one and many, in ways that defy understanding.

This hardly gets close to the condition of 'being RT's body', to the mystery of the 'em' in embodiment. RT's being RT's body was more, much more, than this. Let us count some of the ways.

RT *used* his body as the servant of an agency whose chief and primary agents were his hands. Gripping, pushing and shoving are primordial examples but there were a million others including, most egregiously, his exploitation of the material properties of his body or parts of it: utilizing its weight by sitting on a suitcase to close it; shading his eyes from the now extinguished sun with the hands; or using his head to play peep-bo with a child. The highways and byways of his bodily agency – directly, or mediated via tools and speech – were legion.

RT had *enjoyed* this body. The delicious pleasure of stretching, of scratching an itch, of feeling the sunlight or a cooling breeze on his arms, a warm fire on his legs, of drying his hands on a fresh towel, of giving way to sleep, of a conclusive bowel action, and even more intimate carnal joys, solitary and shared, had been remarkably constant throughout his life. And he had *suffered* this body's vicissitudes. Cold, heat, lack of food and water, injury, fatigue, illness, had reminded him that his subjectivity was the tyrannized subject of the embodiment on which it depended. He became, from time to time, a cold body, an overheated body, a hunger, a thirst, bodily damage, exhaustion. As his final illness had advanced, so he regressed from having an illness to being had by it, to being the surface manifestations of the various quarrels of a body with the world or with itself.

The item in the bed had had other relationships with itself. There had been *ownership* and its modes were many and various. Alive, he had often meditated on the different ways in which his eyes, his mouth, his hands, his shoulders, his genitalia, the small of his back, his heart, his spleen, and his cerebro-spinal fluid were and were not his, and the different circumstances under which they came into ownership. 'My' was differently attached to 'his' face, hand, belly, anus, and spleen. This item had also *known* itself, its possibilities and limitations, in many different ways. It had seen its appearance replicated in photographs and accidentally portrayed in puddles, teapots, and mirrors (car, bedroom, public toilet). (Once he had caught sight of his head on the band of gold now removed from his finger: a pin-sized nipple on his shoulders, next to a spark of light harvested from the day outside.) And, most remote from self-presence, he had been instructed as to his nature through general anatomy, physiology, and biochemistry. RT's factual knowledge of his body encompassed: his weight (the earliest predicate after his sex and date of birth); the origin of the scar on his foot; and the means by which the stickiness of his platelets could be modulated. And there was *judgement*: his estimate of how his body looked, through the eyes of a variety of General Others, a body that would present him as smart, beautiful, stupid, silly, ugly, or none of the above. And *care*: this body had looked after itself in so many different ways; from grooming to preventive medicine, from washing his face to consenting to an elective operation, from putting on warm clothes, to modifying his diet.

We circumnavigate RT's body, trying to awaken ourselves to the relations, now collapsed, that he/I once had to it. Now that it is a corpse, there is nothing that it is *like* for this body to be. It is no longer suffused with agency, or the sense of it. It does not suffer (or enjoy) its states. It does not know itself. It has no owner: it does not deserve the prefixed apostrophe *ess* in 'RT's corpse'. Lacking an appearance that appears to itself, it is beyond shame or pride. It cares not for itself.

Because he is nowhere, *it* is absolutely here – free of distraction into details and the permanent dribble into the next moment that had characterized his living daylight. Stone-cold, it matches the temperature of the surrounding air, no longer a cabin of just-right warmth resisting the chill, in order that its various clockworks might intermesh perfectly, clockworks that had allowed him to stand upright and shake a fellow human being by the hand and put him at his ease. And it is still, with a perfect stillness that makes everything said around it seem contingent, mere chatter.

The visitors have visited an absence: absent life, absent self, absent world. They leave but not before they cradle what was once his head for the last time and give him a last kiss, who, no longer a being-in-a-world, is beyond the reach of kisses.

PART TWO | Before

A Being in the World

three

Organic Accounts

It had served RT well, this inedible deadstock laid out for our inspection, toes turned up, eyes closed, mouth shut, and chest permanently on hold between inspiration and expiration. It had worked tirelessly and, until yesterday, had never had a day off.

RT can impress himself with a few calculations. A lifelong lover of numbers he folds his arms and does the sums, based on an assumption that this body has come to rest after seventy-five years of wandering and fidgeting over the surface of the earth.

The heart, both familiar and alien, inseparable from his most intimate self, beating out a personal time in an impersonal universe, and yet just another piece of meat, had drawn a dotted line between his beginning and his end. This fist of muscle, clenching and unclenching, the motor driving his 5.5 litres of blood to irrigate and nourish the tissues of his body, had repeated its cycle of diastole and systole nearly 3 billion times. The 500 million breaths of a body no longer able to get its breath back would – at a tidal volume of half a litre – have sufficed to inflate twenty standard sized Zeppelins. The gastro-intestinal tract had accommodated approximately 50,000 kilograms or 50 metric tonnes of nutriment, some 700 times his adult body weight. The colon had prepared and delivered itself of 15,000 kilograms of faeces, and the kidneys had produced over 40,000 litres of a cordial whose ingredients had been continuously adjusted with exquisite precision to maintain the stability of the polyphasic system that was the going concern known as RT. Such corporeal bookkeeping was astonishing: the water fell but the waterfall had stayed still. The unobtrusive seepage of saliva would, if harvested, have filled a 6,000 gallon tank. Over the course of 200 million blinks, most of them involuntary, the organism had spread nearly sixty litres of naturally occurring window cleaner across the cornea so that, clear to the end, they had seen him out. Less visibly, through the heat exchange that had made it cosy to cosy up to him, he had, via conduction, convection, radiation, and the evaporation of sweat (in order to maintain its temperature at exactly the right point between absolute zero and that of the interior of the sun), donated something of the order of 30 million calories to the outside world.

Thus did the organism RT – unreflectingly hewing to the processes, mechanisms, and activities necessary to the fundamental goal of survival – inadvertently, if not always effortlessly, accrue astounding totals. For the most part, he did not *do* the organic activity that had added up to such sums in his life. Of course, he spent much time looking after his body, in a manner that was inseparable from his looking after himself, as if in some important sense he *was*, after all, his body. But this depended for the most part on his body looking after itself – not only in the double-entry (food and water, and air) and double-exit (faeces and urine) bookkeeping sense we have already spoken of but also in a million subtler ways. The precise control of the internal environment, with hundreds of parameters being regulated – serum potassium, sodium, magnesium, oncotic pressure, to name a few at random – ensured that the tissues of which he was composed were able to function. Yet more subtle was the orchestration of the chemical activities inside the individual cells, an extraordinary dialectic of soup and scaffolding, of signals and receptor sites, of enzymes and their substrates.

There had been times when RT had tried to stand back from his body and see it for the constellation of miracles that it was. He seized hold of it through metaphors and the one that lay most conveniently to his idle hand was that of the machine. And even this lazy cliché could not conceal his astonishing nature from himself. He saw machines – the organs working together; machines within the machines – the multitude of mechanisms within the organs; machines within the machines within the machines – the pathways and exchanges and systems within individual cells – until under the sharpening acuity of his gaze the scaffolding gave way to soup, a soup which, nonetheless, provided an even more intimate scaffolding teased out in those beautiful biochemical equations, those biologic machines that connected inputs with outputs at the most basic and microscopic level. The idea of RT-as-machine failed at this point. And so it should; because machines are made for specific purposes by artificers other than themselves. No one made RT with any purpose in mind.

Nevertheless, his body was the meeting place of a multiplicity of mechanisms that had to withstand the contingencies of a universe that did not have him in mind: the stresses of cold, of heat, of predators, of thirst and famine, and injury. And there was the necessity for self-repair, conducted while the daily business of survival continued. On the sole of his foot was a memorial to an incident, never forgotten throughout his life, when at the age of six he trod on broken glass in a swimming pool and severed an artery. The gathering of fibroblasts had formed a scar that was able to grow with his growing foot, as he changed from being the terrified little boy being stitched up to a doctor repairing other people's children: a miracle of tissue bioengineering. This was but one of the campaign medals collected in his lifelong war against dissolution. The one example that would not be available to his reflection was the small bruise – on the back of his hand, where the cannula had been placed to allow the drip to irrigate his dying body for the last time. This commonplace purple patch betrayed nothing of the polyphonic conversations between sophisticated molecular mechanisms that had detected the damage to vessel walls and initiated the final, if futile, attempts at restoration.

But circling round this body with an iPad in one hand (giving access to Wikipedia) and a pocket calculator in the other bypasses the central mystery: the distance or distances, evident even in the final days, between 'him' and 'it', between the life of the person and the mechanisms of the organism, between RT and RT's body, innocent of the ownership signified in the apostrophe ess. Of more relevance to this than the stamina of the automata working for his continuation were the deliberate actions they permitted. More astonishing than the volume of breath entering and leaving the lungs was the appropriation of that breath to the shaping of shades of meaning, the transmission and reception of communicated sense by which the deceased had seemed to grasp part of the world, acted upon it, and had been acted upon by it. More impressive than his beating heart was his decision to go to the doctor to have said heart 'checked out'. More pertinent than their passage through his body had been the journeys taken by food and water converging on him from the four quarters of the earth, from distant Brazil and the corner shop nearby, and afterwards diverging from him to equally distant parts.

His slim frame had been the waist of a chiasmus; or (more aptly perhaps given that the passage of time comes to know itself in him and others of his ilk) a kind of hourglass through which a future had come to him from many angles, travelled through the present, into an equally divergent past.

This body frozen between inspiration and expiration gives no hint of the panting, laughing, crying, belching, sighing, soup-cooling, speaking (whispering, shouting, muttering) creature imposing his meanings on the world that imposed meanings on him. And there is little in these remains to suggest the upright, prehending, and apprehending, articulating man acting and reacting; even less the thinking (forethinking, remembering, puzzling, planning and plotting) 'he' who had shaped the world that shaped him.

Inevitably, the search for RT within RT's body becomes a search for the gaze that his eyes once had, that, in virtue of drinking the elixir of light, had mysteriously situated him in great volumes of space. Space – indeed spaces – centred on him that were now invisible. Spaces in which this now inert carcass had stood up and walked towards destinations and goals, aware of the dialogue between his feet, his shoes, and the surfaces on which they had walked; into which RT had reached to palpate, grasp and manipulate worlds, both directly and also indirectly through countless instruments and proxies. Spaces where he had met with others and parted from them, where he could say hello and wave goodbye. Spaces that had been vastly supplemented by the semantic spaces he had breathed into being. Spaces that had upheld 'here', and 'there' and 'elsewhere' and have now collapsed.

In pursuit of these spaces, we leave this room and enter the world he has left.

four

Elements

There is no right way or right place to begin when you are trying to look at life from without, when you are trying to grasp a world. A sphere begins where you begin. All we can do is try to start with the most familiar of the familiars he might mourn if he were capable of mourning: those elements that surrounded him all his life, and indeed preceded his entry into life; the immemorial givens transformed by the human world of which he had been a part. They forged, sustained, threatened, and ultimately reclaimed him.

Earth

This is the elemental element, the ur-stuff, the ground of his being. It shares its name with the planet on which he – and all the hes and shes that there have ever been – had passed his life. His weight, the most continuous and ubiquitous manifestation of his carnal self-presence, had been a manifestation of the earth clamping him to its surface, a force mitigated by carers and machines.

Driven by appetites and desires, fears and hopes, needs and dreams, duties and whims, near and distant goals, he had travelled over, or close to, the surface of this massive ball of soil. In defiance of the mechanical instability of the jointed altitude of his body, he had stood up and walked. Walking, he experienced the different manifestations of 'the ground', the 'under-foot': bouncy turf, glutinous mud, crippling rubble, slithery sand, treacherous ice-glazed pavements, delicious firm paths and foot-chafing tarmac.

Earth was the stuff that his forebears had transformed into land, property that was owned and worked, hedged, walled, fenced, fortified, dug, sown, grazed, built on, fought over, concealed, opened up, mined, and shaped into artefacts (pots to hold food and water, dwellings to hold lives). When human lives began to be devoted to wooing fertility out of the earth and the earth became The Good Earth, mankind developed a collective memory and pre-history passed into the history in which he had been a recent participant.

The Good Earth was the ground of his being but also that from which he distanced himself. Earth was dirt; and soil soiled. It was washed off his mucky face, laundered out of his clothes, and swept out of his dwellings. Mingling with the air it became a kind of drizzle, a coarse smoke without fire, motes that blocked his nostrils, made him cough, and dried his hands so that his fingers disliked the feel of each other. It formed films over lintels and furniture, attached itself to books testifying to their being antique and neglected. In order that his acquaintance with dust might be only brushing, he belonged to a race that knew the verb 'to dust'.

Ultimately, this had been in vain. His human clay, that signalled its nature when it soiled itself, was destined to be returned to its source. The dirty face was always en route to being effaced to faceless dirt, to joining the rubble of the ontological rabble.

Air

For most of its history, the air of planet Earth was unbreathed. Then it was aspirated by plants, subsequently by animals, including those underwater creatures whose gills could extract oxygen from the drowning places of lung-bearers, most recently by humans, and finally by RT. His life and breath were co-terminous.

The restlessness of the world around him was broadcast in the sound of the wind, the spume blown off waves, the grass shivering and trees bending so they would not break. He glimpsed the coherence of this restless universe in the connections between the flight of clouds, the trembling of foliage, the nodding of small flowers keeping their heads down as chaos raged, the guttering of a candle-flame, and the billowing of a skirt. Gales, breeziness, and stealthy currents, barely perceptible agitations too weak even to extract speech from a poplar tree – he had known them all. He had seen dust animated to tassels of false smoke, crisp packets ape tumbleweed, gusts slamming doors against themselves, curtains grow gravid, and paperweights justify their existence. Air that on a summer's evening was unable to discommode thistledown could gather up its collective strength and, revved up from zephyrs to breezes, from breezes to gales, from gales to hurricanes, smash everything in its path, uprooting forest trees like weeds, clouting houses flat, and driving the sea inland in search of victims to drown.

He knew so many ways of wooing the wind and keeping it in the place it should have in his life. Fans, open windows, air conditioning, ventilators: all freshened air that had become too hot or polluted. And, outside, for business or recreation, he could cheat the wind in many other ways than by wearing the eponymous article of clothing. Sails, kites, paper planes had underlined what intrinsic delight there was to be found in the unconstrained boisterousness of the air, which, in high places, prompted him to dream of ticketless aerial flight that required of him only that he stretched his arms and let go.

Like many living organisms, he had been a lifelong windbag. He had been aware only intermittently of the inflation and deflation accompanying him from morning to night and indeed throughout the night – as when he paid back an oxygen debt in the silence after vigorous exercise, or he opened a window and drew in the sweet air of the summer morning, relishing the lungfuls and the luxury of expanding his chest. More often, he heard his own voice, that miracle whereby the gas exchange necessary for physical survival was transformed into information exchange necessary for social survival. He had marvelled at the variety of ways (sighing, screaming, crying, and laughing, as well as talking) in which the treasure of his chest was unpacked into strings, rags, boluses and tapestries of meaning that he sometimes dismissed as 'gassing'. Whether or not the universe began with The Word, with meant meaning, the word was close to his beginning and his end.

Water

He had been an oddly shaped, largely fluid-filled, sachet, living on a planet 72 per cent of which was covered with water. From the circulation of blood and lymph, upon which all else was predicated, to the intimacy of the saliva welling up in his mouth, and the cellophane of sweat clasping his head; to the glass of water, a quenching cylinder of liquid slaking the darkness within, and the reassuring splash as he rushed into an unlit room in response to a call of nature; to damp patches on clothes and walls, and pools registering a shower with a multitude of startled eyes, and overflowing gutters speaking one of a thousand aquatic dialects; and to streams, cataracts, and oceans. From the amniotic fluid of the months when he was prepared for life on earth, to the final hours when he drowned in his own secretions and nature reclaimed him, it was water, water everywhere.

While water has no memory, it left many memories during a life in which this fluid coursed through every corner of it.

Rain had fascinated him when it did not depress his spirits. Those widening circles of water travelling over water, intersecting with inexpressible grace, were a beautiful expression of miraculous symmetries in nature, usually hidden from his gaze, symmetries that forbade the ripples to form enlarging squares or ellipses. He had been entranced by the effortless artistry of pools, creating likenesses of the world around them and changing their style of portraiture according to the weather, or marking the passage of a paper boat with a corduroy of interlocking Vs attached to its stern.

Water was as polyphonic as the air: splashes, gurglings, the monosyllabic plink of fat drops dispensed singly from the end of pipes and branches, the fine-grained splattering of the fountain or watering can, the poster-brush-wide organ roar of the waterfall, and the colloquies of the rain with the puddles it created, of the falling with the fallen. Water had come in a variety of portions, from the horizon-encircled to the porcelain-constrained, served up in lakes and leaks, in seas viewed from ships made tiny by them, vast and violent shores observed from cliffs, reservoirs that took a day to circumnavigate, streams he could leap across, pools on the pavement, glassfuls angling the light, thimblefuls, runnels of tears coursing a face, drops that took a full minute to become corpulent enough to earn their escape from the end of a twig. He encountered this transparent, chilly, elusive stuff in measures that ranged from ocean-deep to the two-dimensional slimness of a stain.

The water that had flowed through his days was for the most part tamed, though uninvited rain prompted him to run for shelter and (once) a life-threatening rip tide reminded him that his hegemony over this and the other elements was only temporary. Temporary, yes; nonetheless, it was impressive how the mindless power of water, obliged to seek out the lowest accessible point, could be subordinated to his most precise intentions and his most carefully calibrated needs. The regimentation of rainwater, via reservoirs, into tap water, the disciplining of boundless, edgeless stuff into discrete aliquots delivered at the time and to the place of need was a signal triumph of man-made order over natural insouciance, of his conspecific's unique capacity to suborn the general habits of the material universe to singular human requirements. The circuit diagram of pipes dragooning a silky formlessness to a pattern connecting thirsts with their quenchings – the unfelt thirsts of cultivated plants and the felt thirsts of animals and the spontaneous and regulated thirsts of humans – was a triumph of the collective human mind. As, too, was the transformation of water into a utility, a saleable commodity, an ingredient of processes and of materials.

The containers required to make water portable and storable, biddable and potable, were almost as ubiquitous as the stuff itself. Cups, buckets, watering cans (spouting minced liquidity, gargoyling through a multitude of stomata), tanks, wells, reservoirs – these placed water on permanent standby, ensuring that it was always on tap. Stream to reservoir, reservoir to pail, pail to jug, jug to glass, and glass to mouth: by these mediators had a torrent hundreds of miles away been directed down his throat (now no longer parched) and a reservoir been served up in swigs and sips.

The intimacy and frequency of his many dealings with water was a consequence of the fact that maintaining inner-fluid balance was one of the chief conditions of his continuing existence. The components of the various inner, carnal seas that, ultimately, bathed the cells of his body had to be maintained with exquisite precision so that the million-cogged chemical clockwork of his body could continue to make it possible for him to be a sentient habitué of daylight, rather than carrion awaiting disintegration.

Water had been requisitioned in a less fundamental way to supporting the task of upholding the distance between his body and the world in which he had to immerse himself in order to live. Assisted by flannels and towels, soaps and creams, brushes, he had groomed the surfaces and nooks and crannies of his body, separating face from dirt, armpits from sweat, and a thousand allotropes of filth from limbs, orifices and clothes so that he could proceed through the world uninfected, rotting only at the acceptable pace, nearly odourless, and smartly attired, keeping the earth's maw at bay. The distances travelled between the rain showers that happened involuntarily and the deliberate act of taking a shower (a delicious focal, elective, warm rain) – so that water could be appropriated as the internal accusative of the verb 'to shower' – were long and sinuous. Unwoven from a lake big enough to accommodate a life-size portrait of a fell-side on its breast, the threads of water emerged as a loose formation of parallel rods, a bespoke lucent spaghetti combed out straight, spun and cut off to order at the beginning and end of ablutions.

His encounters with water – this magic stuff into which he could plunge headfirst without injury – went beyond the utilitarian, beyond the swabbing, rinsing (washing off the consequence of washing), moistening, and drying that occupied so much of his life. Its wonder had first struck him anew when bathing his infant children: they had whopped the surface of the bathwater to create tantalizingly brief chandeliers of drops. Seen from the complacency of waterproofs, or through a window seething with wriggling, translucent tadpoles, even unwelcome rain might seem beautiful. As, sometimes, did glistenings – slimmest of pools – that coated so many of the surfaces he walked upon or between – and the puddles in which he enjoyed the minor transgression of splashing. The spectacles of the steaming pavements, of wetness unpacking itself to smoke, of a granulated mist with elusive edges, drops finer than dust, entranced him with their ambiguous ontology: poised between being and not being, available to the eye but not to the hand.

And he had relished, finally, how cold conferred rigidity on this, the very paradigm of fluidity, making a portable window out of a puddle, fastening a monocle on to a tarn, arresting raindrops to zero-calorie sweets, transforming showers to feathers fallen from a giant bird or from a pillow fight between clouds, asterisks dictating hush, and organizing window-mist into fern prints.

Fire

In its purest manifestation, flame, it was fluid, insubstantial enough to be pushed around by the air, opaque and yet a pure visibility, turning the air in its vicinity into ripple glass. It was Becoming almost completely liberated from Being so that even smoke might seem solid by comparison.

His relationship with fire, as with the other elements, was ambivalent. A remote child of the sun, he had needed protection against the radiance of this ultimate ancestor but could not live without his portion of it. The low flame of his metabolism was not enough to keep him warm, even when his bodily heat was defended against dissipation by clothes and dwellings or by largely symbolic pyrogenic activities such as rubbing together those hands that now can no longer clasp each other, each taking warmth from its enantiomer, or harvesting niggardly thermal benefaction from his pockets.

His greatest good fortune was that his arrival on the planet had been postponed until the species to which he belonged had learned the art of controlling and rendering portable the hitherto uncontrolled and uninvited wildfires that sometimes consumed forests and villages, cancelling in a few minutes centuries of growth, and years of life-husbandry. And so he had taken it for granted that it was possible to harness this seemingly unbiddable element to cancelling cold, banishing darkness, creating the materials out of which artefacts might be forged, and transforming the fruits of the earth into something tastier and more nutritious. By the time he had entered the world, fire could be struck into life, nursed, switched on and off, and regulated with exquisite precision.

Many of his deepest, or darkest, or most inward memories were lit by fire: the flames lighting the trees round the camp, the reflection of the glowing coals in the Christmas baubles, the celebration of the power of fire in fireworks. And his acquaintance with so many modes of being cold – numb-fingered, frozen-footed, chilled to the bone – had been compensated for by a proportionate number of ways of being cosy and warm, with the best of all heaters, the body of another. Snugness had been one of the great unspoken pleasures of the life of this body, now cold as the meat in a fridge.

Cycles of warmth and cold marked the larger periods of his life. Cool mornings gave way to sultry afternoons, and hot afternoons to cold nights; the succession of days forged a connection between chilly winter and hot summer, or the disappointed expectation of it. And warmth and cold were separated in space as well as time: in the hot tap distinct from the cold, the warm hearth from the draughty door, or the bitterly cold Northern forest from the stifling Mediterranean street. The cycle of the seasons was matched by a cycle of distinctive livery: wrapping up warm versus stripping off, sunbathing and sun-lotion; the autumn collection, short sleeves, thermal underwear and swimming trunks. Such were some of the consequences of the tilting of his planet.

The very real and yet elusive existence of flames had fascinated him as a child, and the fascination endured. They were tethered rags of magical fabric, unwearable shirts, illuminating a darkness broken into a million shadows made to dance to their endless fidgeting. Only slightly less elusive was the smoke that had crept round corners and through crevices, and seeped through the sides of his life. Plumes had stretched in the sunlight, as if enjoying the warmth of the sun, a squirrel's tail unfolding its blue mist in a forest, or shoulders shrugged at the passage of time.

Compounds

Marriages between the elements produced astounding progeny.

Consider those children of air and water: bubbles. A lucid membrane separating air within from air without, they had entranced him with the miracle of their birth and the continuing miracle of their unpopped endurance. The lightest currents of air provided the motor for frailty-defying journeys of lone spheres, perfecting their shape as they broke free of their origin. Closely packed crowds, suds crepitated in the washing up bowl or (more *sotto voce*) in sponge-like heads of beer. Sparkling mineral water conversed with itself, having announced its nature with a sneeze as the cap was removed from the bottle, permitting the pockets of trapped air to inflate, to escape and to swarm lemming-like to extinction, or translate their demise into a Velcro texture of pinpricks in the mouth.

The relationship between water and earth was riven with pleasing paradoxes. Together they formed the primordial filth: mud. And yet water washed away mud and mud transformed by fire into bricks kept out the rain. The ewer that brought the water to separate the earth-as-muck from the mucky face was itself the product of an interaction between earth and water, directed by fire.

Fire drove water through its phases: the metamorphosis from ice to steam, or to vapour even more translucent than itself. It could make any liquid babble fluent Bubble-speak as it evaporated to steam. Flames made this ubiquitous solvent more ravenous for materials that would otherwise be dissolved slowly or not at all.

Once air had been expropriated by RT's ancestors to emit meanings that had been 13 billion years in the making, and ideas as well as energy could be exchanged, there had been no apparent limit to the uses to which the elements could be put. The air's ebullience was captured by sails for relatively frictionless journeys over water or to power other machinery (windmills, wind turbines) that served purposes hitherto unknown to nature. Irrigated earth was husbanded to utter breakfast cereals and, indirectly, steak dinners cut from the bodies of grazing livestock. Water, contained in earthenware held over fire, became the hot drinks that punctuated, even measured out, his days. The transformation of water by flames was harnessed to the self-propelled vehicles that removed some of the trudge out of journeying and enabled work to be delegated to machines.

Thus a glance at the alchemical genius of a creature – *Homo*

Sapiens (HS) – able to transform the elements that surrounded and pervaded him into commodities which made the stuff ('fire', 'air', 'water') into eponymous activity. He had been part of nature, with natural needs, and also apart from nature, meeting those needs in an unnatural way, and acquiring on the way needs unknown to nature.

Light – The Quintessence

And so to light, heat's inseparable companion in fire, present most primordially in the sun which had warmed and lit his days. If fire might have seemed closer to thought than earth, light seemed closer than heat to consciousness.

Light and his life had been inseparable. Not for nothing had his conscious existence been decribed as 'living daylight'. His arc, from non-existence to non-existence, from impotence to potency to impotence, was through brightness bounded by the darkness of the womb to the unfelt darkness of the tomb or the urn. His life had been most naturally divided into days, stretches of light hyphenated by intervals of darkness.

One version of his biography could have been written as a 'chronicle of lost sunlight', patches of light, memory-flares illuminating his vanished world.

The square of sunlight on the nursery floor bathing a teddy bear's leg, the radiant green of leaf-dappled grass, a spark in the eyes of a loved one reflecting in miniature the view she is looking at, the old-looking light in a cupboard opened after a long time shut, the evening sun round the ankles of a hedge elevated above a deep lane leading down to a sea announced in the falls, pulsing aches of sound, of gulls, the platinum primrose-grey light hue of a remote tarn, the yellow of the sun-drenched stonework in the public square on a hot afternoon, the quiet candescence of the last leaves in a twilit November wood, dawn tingeing mountain crags and pinkness on evening clouds that had seemed like added pillows to his childhood bed, the tingling sequins of silver sea light. Gleams on furniture, on polished toenails, on the polished curve of his denuded cranium, sipped from clouds, glimmers caught on distant pools from remote towns, the spilt marmalade of sodium lit tarmac. The lunar glow of a child's face looking at a screen as she texts a friend. Great lights and small lights, public lights and private lights stitched together into mornings, afternoons, and evenings, the shining hours. And the thought, many years later, that these places still exist. That the sunlit square ruled over by a huge plane tree still has a dog walking past a table, another awning being lowered, another leaf falling to the ground.

He had glimpsed himself as a reflection, the protagonist in a patch of light, most arrestingly in windows, where, looking out at the world, he encountered his gazing ghost looking back at himself looking out. Of the many instances, he would, if he were now able, have alighted on his study, early morning, just after dawn, in winter. An intricate black-brown twig-scape. Above the bare hedge, a replica of his desk, computer, hub, printer and paper, and the faint image of the book-filled shelf above (and a square alcove with spare pens, notebooks, roll of sticky labels, souvenirs from holidays). An illuminated page in a dusky room beyond the focus of the anglepoise staring at the text. The page in his hand feeding an inner world. Nearby, the family asleep, remote in their dreams. A silence outside unbroken by birdsong or traffic. He exchanges glances with his image looking at himself. Living in the light.

He had compiled an album of privileged spots preserved out of the million hours of daylight, stills from a hundred miles of footage on the floor of memory's cutting room. Memories of glimpses, of places whither he had taken his head, trawling for qualia. A corner of a cobbled street, a square of lawn in a quadrangle where tree shadows actively dappled, a sunlit bay glimpsed through cricket-mad pine trees, the interior of a library with a marble statue picked out by studio lights. A lamp by the harbour on a hot night, with its network of attendant flies, reflected on the sea, his ultimate experience of openness, giving the false reassurance of a gentle swell. A table-light in a Belgian café picking out dark-panelled walls and a display of artificial flowers. A torch lighting the way through a frozen backyard.

Random snapshots from moments, hours, weeks, openings that snapped or softly closed shut. A dancing dazzle on a beer glass in the café Slavia with Hradčany behind him and Charles's Hunger Wall ahead; glimpses, from the concourse, of oblongs of butter-light sliding down the face of a Pendolino drawing into Euston station; the light under the bedroom door as he tried as a child to sleep; outside light scissored to geometry by a window frame, a trapezium on a rose-patterned carpet; slits of sunshine filtered through the wooden shutters on to a perilous concrete concourse necklacing a tower. And their complements – those shadows that show forth the light: of a poplar tree on a hot summer evening, connecting the edges of a narrow, unbordered single-track road pencilled across a great plain; of a hand making shadowgraphs on a bedroom wall; of the stone penis on the stone thigh of a naked statue in a square in Rome; of a cloud quenching the brightness of a mountain side.

And, between light and dark, the second helping, provided courtesy of the moon, the sun's hollow echo. Moonlight did not banish the night's darkness as much as reveal it, lowlighting its threatening or merely impenetrable complexity. Waxing from nail-paring to full globe, it was the poignant marker of the passage of a fortnight's holiday, from a beginning with a wallet full of free time to a spent up end and the return to the treadmill. Its gleams on rivers and streams or on wet slates seemed the portraits of gleams, quotations from a world of the imagination. The lunar glow in the garden at 3 a.m. was sometimes sickly, likely only to reveal predatory creatures and their terrified prey, and underline the distance of the night hours, of too-alert insomnia, from the longed-for ordinariness of the daylight world. It was the viewpoint of septuagenarian solitude looking back at the summer's days spent on the beach with the children when they were small.

Light – and dark. Shadows faithful to the outlines of objects, pure forms, negatives, tethered to the feet of things incised in the light. The 'welcome' shade beneath a tree, broken into soft-edged pencil sketches of leaves, and the dunnock-twitched interstices of hedges. The solid darkness into which he was plunged when the power failed, the pocket-sized blackness under the blankets or in a hiding place, the great darkness of light-excluding forests and countywide starless nights. All the dims and dusks, the drawings in, the crepuscular fringes, giving warning of the sunless (and moonless, darkless, dayless, nightless) zone to come.

There had been some thirty thousand days between his first breath and his last, each shaped according to the crescendo, the changing angle, and the diminuendo of light that had shortened and lengthened the shadow that he, being opaque, had cast. The patchiness of memory had divided the continuum of passing time into distinct territories, each with its own ambience, its own moods, its informal checkering of brightness and gloom: dawn, noon, and evening. It is these that he would wish to acknowledge in any ceremony of fare-well: a dawn in late autumn, an evening in spring, a hot afternoon in summer, and a winter morning.

five

Space: Senses

So much for the elements, the basic stuff out of which he and his days, the world, natural and transformed, were made. Of course the air is not in itself crisp or scented or vocal, nor water wet or shiny or splashy, nor the earth mucky, heavy, or obdurate, nor fire hot or smoky, or light bright, soft, multicoloured or plain. Nor are their manifestations set out in space. They are not near or far, to hand or remote beyond imagining. Something must have opened up the material universe – an insentient continuum – into the world into which he passed his life, allowing heres and theres and elsewheres to blossom. Something slit the silence so that the world could sound out, broke the darkness permitting the light to illuminate.

That something was RT's now extinguished sensorium. His senses had reached into a space that they in part constructed, in part encountered, in part synthesized, and in part suffered. From them arose – in a universe without any special places, or indeed any places, in an unimaginably large desert of stuff, a pure extended-ness without qualities, indexed, where there is no one, to nowhere and no time – places in which he was at home and homeless. Here, there, and elsewhere in an everywhere that is in itself nowhere.

It is to this senseless universe that he has now returned, as a disintegrating part with no particular distinction. His cold frame on a cold bed is not cold for itself; nor does it have the 'on' linking it to the bed. Those qualities and relations lie at the heart of the distinction between the 'it', which does not even possess the space it occupies and is, and the 'he' who once was 'me'; between the corpse that merely 'is' and the 'I' who 'am'. For the worldless corpse there is not even emptiness, nor darkness, silence, or insipidity.

Let us glance at what has been lost; at the senses that transformed into exposure the interaction of the flesh, that went under his name, with its surroundings. These opened up a space that he enjoyed and suffered, a space in which he controlled and in which he was helpless, in which he was located, and where he located, and endlessly relocated, himself and the other contents of his world. A space in which he was a being 'here' in a world that was 'there'.

Looking and Seeing

The empty gaze on which stiffened lids have dropped the final curtain is the most potent sign of his absence from the world and of the world's absence from him. The gaze fulfils the idea of a person as a viewpoint at its least metaphorical and it illustrates how the viewpoint is, potentially, a vantage point. The eyes of upright man are elevated by being located in a head held aloft. A scene, packed with objects, swarming with events, dense with significance, and rich with possibility, is revealed to the gazer. He looks vacantly or merely gawps; or he stares, peers, peeps, scrutinizes, judges. He pans round, at home and abroad, and is taken out of himself, as successive slices of the world, or light cones, are transformed from beings to revelations. The gift implicit in The Given was most clearly manifested in the world as spectacle.

And he transported his head to places where he could extend his visual field. It had been an early joy to climb a tree – one of the most ancient of natural look-out points – and, from a vantage point rocking in the wind that enabled him to come closer to the freedom of the untethered clouds and promised make him part of the sky, to look down on the garden, and the roof under which he passed his childhood. Or to look out of a high window at the people and cars and houses below. Or to be on the top deck of a bus, watching milling passengers and to be granted the temporary privilege of looking down on the crowd rather than being part of it. To be elevated above a road bending out of sight was to take a sip of the gigantic elsewhere in which his Here and There were located.

A larger draught was available when he observed the houses in geometrical rows from an open field modestly rising above them, a few miles from his childhood home. Standing on a rocky outcrop, or a mountain ridge, and switching between the totality laid out before him and minor details – a blossoming hawthorn tree by a farmyard, a sheep loosened from the flock, a motorbike and a car converging on a crossroads, a single cloud across an otherwise cloudless sky, or curtains of rain disguised as lines of darkness – gave him a compendious sense of viewpoint. So much was reeled in with his gaze that had sipped the knowing wink and gulped the mountain range and all between.

Vision had been, of all his modes of perception, the foremost bearer of space. By bringing together as an ensemble, or a field, items that in themselves carried no awareness of each other or of themselves, RT was in part the co-creator of the space in which he found himself. His gaze made the world most immediately and extendedly present and gathered its presentness into an incomplete sphere.

And the gaze was itself visible. He saw others, saw that they were looking, and saw the eyes with which they looked. Nothing comparable was true of hearing or any of the other modes of perception: there was nothing audible about the process of hearing. Which was why, of all the senses, vision most explicitly joined his attention to that of others, and the space into which his eyes reached was most explicitly a shared space.

When gazes are exchanged, so that each one looks into the eyes of the other, a different kind of space is created. This was the space of the community of human minds to which he had once belonged. With this community came a new mode of looking, one that looked back into histories and memories, shared and unshared, and to spaces that lay beyond the bounds of vision. Looking was shadowed by thinking, light by words, and the seen world coloured by emotions that in part narrated themselves. In a reflective moment – as when he looked up from a book at a tree swaying in the sunlight – his gaze was here and not here in a way that only he could have known, connected with his immediate surroundings and at the same time concerned with worlds drawn from the past and the imagined future, from elsewheres remote from this over here and that over there.

The scene, the landscape, asked to be read as well as registered as visibilia, to be interpreted. But every now and then, in moments that he liked to recall in hours of darkness, the seen was pure: red was an irreducible essence, colours were flames, and the yellow of a daffodil was a solar trumpet and the crystal blue even of blue eyes was simply blue.

Hearing and Overhearing

The pinnae on the side of his head seem more than ever clipped on as an afterthought now that the lobes are blue with stagnant cyanosis, creased with the unrelieved pressure of lying on his side in a few moments before and after his death, and no longer cocked to a silence pregnant with messages. Sounds joined with light in establishing the exquisitely folded layers of here, there, and elsewhere.

Listen.

Closest, the humming of head that grew more obtrusive with the years, a steady downpour of acoustic filings, dissolving seltzer, murmurations of midges. The crunch of nuts inside his mouth, or the gravel-path vocalizations of fresh celery, or the borborygmi in his belly. Next, clothes-talk: rustle of silk, the high-pitched whish of something being brushed off, the slightly surprised cry, even protest, of a zip as the formalized tear in cloth, a dentate rip, is opened, the struck-match call of stockinged thighs being uncrossed, the tick, tick, tick of the plastic-coated ends of untied slipper laces, the whiskering of corduroy rubbing its exquisitely tiny furrows, reminding him, when he had a mind, of a man walking up a cobbled street in Prague, the dialogue between waterproofed trouser legs highsounding the arresting quiet of that beautiful city.

Beyond this, the near abroad. A chair made to squeak by someone's restlessness. The soft fingerfalls of keys on keyboard. The pencil writing quickly over the page in a very quiet room, sounding like an elfin panting. The throbbing of whatever gadgetry it is that keeps the place he is sitting in warm and dry. Further out, a tangle of voices, with occasional phrases audible, like flowers half-hidden in undergrowth. The clatter of plates as someone whose turn it is washes up. Children's voices – in the next street, in the neighbour's garden, in the town square, on the beach. Outside, the traffic: all those intentions picking their way through the tangles of others' intentions. To hear them with ears unwaxed, syringed by the idea of the eternal unbroken silence, is truly to hear them.

The bustle that helped to make 'here' here provided a representation of the world that was more episodic, random, than the continuities presented to his eye: intermittent hissing, scrawpings, chirruping, crashes, murmurings; birdsong, sighs, snores, squeals of outrage, a splutter of laughter, tyres' hyphenated passage over damp cobbles, the simmering of rain on a pool or of a kettle boiling, the rustle of leaves, gravel provoked into utterance by the postman's bicycle, the roar of a lion or a waterfall, the slap of the sea against the bow of a boat tethered to a harbour wall, the teasing sound of something twitching in a hedgerow, the sound of the word 'twitch', the sound of the difference between these sounds, the sound of an oncoming train, of the zipping and unzipping marking the beginning and end of lovemaking, their similarity discounting the interval that heard the broken song of approaching orgasm, and further out – the ticking clock in the hall, which in the small hours seems to convey the menace of secateurs pruning the hours, a dripping gutter, an overhead plane. The modes of scraping: the nail against the blackboard, the frost off the windscreen, the mud off the shoes, and the dog trying to enter the house.

The incidental pleasures of sounds. Noting natural homophones such as the distant bruit of heavy traffic imitating a pine forest animated by the wind, and both ventriloquizing the sea breaking on the shore. Or a sun-lotion bottle squeezed uttering a sound somewhere between a wet fart and a cat scalded by the unexpected presence of a rival. And echoes: multiplication, mockery, acoustic fuzz in empty space making surfaces into sounding boards. Deliberate, accidental, natural, artefactual, empty, charged with meaning, musical, dissonant. The joy of near and far brought together: the quiet rasping of his scratched beard sharing air-time with a fox's distant call, gulls unfurling marine spaces mingled with someone's breathing made audible by turbulence on the turbinates, the steady blizzard of sound fragments inside his head with the barely heard music carried on the wind from a distant fairground or the coarse-grained hail.

The domestic voices: the sounds of cleaning, with the vacuum bellowing like an animal in labour, the hot breath of the hair-dryer, the whispering of dusters dusting away dust. The bass register of the yard brush, the tenor of the polisher, the anguished soprano of the wet floor cloth. Scales of sound: broom, clothes brush, toothbrush (that changes its timbre in the echo chamber of the mouth as attention moves from back to front teeth). The tinkling of spoons against porcelain, the slam or creak of a door. The many utterances of paper: flapping to the command of fidgeting air, the yelp of the torn sheet, and crackling, rebellious stretching as paper balls unscrunch in the waste paper basket. Clicks: of a switch, of a cap on pen, of a sneck opening, and a box closing. The asynchronous chorus of artefacts into which housework, fragmented into discrete tasks, was outsourced. Washing machines masticating clothes, central heating pipes cracking their joints, electric kettles torturing water, gas fires snoring, lawn mowers murmuring with approval at the neatness of their handiwork, close-shaving the grass, the discreet hum of the computer hard drive – all punctuating their work with progress reports coded in beeps, bips, clicks and ticks.

They had added their incidental voices to the invited voices of radio, television, and the music centre. And there had been the sounds announcing the encroachment of the outside: the knock at the window, the doorbell, the newspaper being wiggled into the letter box, the car coming up the drive – thus the wider world seeping into his small parish of ownership.

Special sounds. Voices, and music, of course. But also footsteps. He had often relished his solitude by listening to his feet cutting quietness in two as he ate into space, while hearing the faint wind in his ears, a primordial sound heard by bipeds and quadrupeds alike. There had been so many dialects of foot-speech: the hurried successive clips of a woman on high heels, the trudge of a booted male, the patter of a barefoot toddler, the blurred, soft shuffle of the slippered elder, a smeared cadence of ambulation no longer crisply divided into discrete units.

The hands were almost as various in their vocalizations. In addition to rubbing (accent varying according to the dryness or wetness, the hardness or softness of the skin), tapping, and finger snapping, there was clapping, that served so many different purposes: signaling a rhythm to be followed, or an attention demanded, or approval or delight. His had often joined with others' percussive hands to generate an appreciation that imitated a fat-dropped downpour of rain, or a waterfall, or, when there was more politeness than enthusiasm, waves on a quiet shore withdrawing over pebbles.

In the comparative peace of the domestic front, he had sometime silently rehearsed certain special sounds from the world at large. At random: the howling of a winter gale that underlined the stillness and smallness within, desolation making smug comfort more comfortable; an April blackbird defining spring spaces from the coigns and pinnacles of the neighbourhood, with lowly passerine cheeps pointing the joins in the spaces opened up by the songs of the songsters; beyond the city, a skylark climbing up its own raga, summarizing the summer miles between cliff and moorland; and church bells on a warm September evening, transformed by memory into a perfect gathering of hedges, clouds, cottages, and childhood hours, as if they had been the voice of a landscape composed of especially precious intersections between patches of space and patches of time.

There were almost as many modes of overhearing as of hearing: down stairs from upstairs, through an open window, over a fence or a hedge. The distant, angry car horn invited him to overhear his own thoughts and to think of the ships' horns on the foggy Mersey of his childhood, of hunting horns, of the melancholy Last Post mourning the dead, and the utterly beautiful calls of a French horn exercising its voice in a concerto.

And then there was the unheard, the silence that seemed at times like soft ear-balm, picked out and shaped by sounds. It was present of course in music; and in speech, where without the scissoring of conversations into utterances, utterances into sentences, and sentences into words, all separated by silence, there would be no meaning. But it was also there in escapes, calms, and pauses, as when the dropping of water from the trees after the rainstorm celebrated both the present stillness and the wildness that had preceded it. And there were moments when sound and silence were in perfect equilibrium, as when he had listened through a half-open church door, and had heard music – closest to the susurrus of wind brushed by the pines, the Aeolian harp of the telegraph wires – speaking, to a non-existent God, of our lives between soil and sunlight.

Thus the overheard comings and goings, the drippings and bangings, the music and the noise, the chorus of the trees, the birds, the feet, the mouth, woven into the ragged tapestry of sound that accompanied his days. Until the day when all signals dissolved in noise and both signal and noise were silent in the head that had once been his. Never again will a scrawped chair or a banged door or tuneless whistling irritate him; or loud bangs startle him, or fountains soothe him. The static on the untuned radio, its ear not quite pricked to harvest incoming meaning, the crackle on the vinyl, the frying silence on the phone – all gone, along with the words and music with which they once shared acoustic space.

Sense Through Scents

His lost world had been predominantly luminous and sound-filled but it was also odorous. Smell had its equivalent of the staring and peering eye or the cocked ear in the sniffing nose. Reverse nose-fluting, made smelling an activity as well as a property. The intelligence it yielded was intimated rather than stated. Smells were often pervasive, untethered to any evident source, the very essence of atmosphere, more like adjectival modifications of the air than the referents of the nouns that named them. There was no olfactory space analogous to the visual field, or the acoustic space built up out of the places and objects from which sounds arose. The over there had to come over, and be, here in order to be smelt, hence its particular invasive power. The verb 'to smell' – corresponding to sniffing and categorizing and passing judgment – therefore sounds more active than it is; after all cheeses can smell. Smells invaded his consciousness and while he could have avoided the rank and vile by pinching his nostrils, there was a limited time for which he could avert his olfactory attention by this means.

Although loss of his sense of smell would not have disabled him as blindness or deafness would have done, anosmic he would have been greatly impoverished. Smells encoded remembered places, things, persons, and even past selves. While aromas, perfumes, scents, fragrances, odours, pongs, stinks and stenches are sensibilia that somehow creep beneath the main action, attendants rather than protagonists, they are rich and various. If he were capable of experiencing the lost world, he would recall: the inexpressible aroma of an early morning, the scent of mown grass, the fragrance of happiness in the pine trees; the must of churches and deserted buildings; the ravishing odours of food; the infinite variety of the smells of burning – wood-smoke (an ethereality built on a will-o'-the-wisp), toast on a winter evening, dust in the electric fire, paraffin heaters, ironed cloth, singed hair; the many odours, bought and secreted, of the beloved, of strangers, of his own body; the sea scent, suspected far inland; even, perhaps, the urinary whiff of public toilets and the oniony air of the changing room.

He had relished without being a connoisseur. He could not attach names to many bouquets, did not have the capacity to translate tastes into adjectives. But rain on a hot pavement, the anticipation-laden scent of a room buzzing with partygoers, cigar smoke in a hotel corridor, the cool, shy aroma of a rose, had been precious beyond price and caused those nostrils, now collapsed, to flare with joyful attention.

As we move inwards, so we close in on the body and lose space. The tastes he tasted did not create a space of their own and all those things he touched kept their own space but did not create their own version of 'over there'. There had been no tactile field to supplement that which is built up by light.

His collapse, however, is more complete and radical than the deflation of the outside upheld by his senses that had made him a viewpoint experiencing a world from a particular place and every-thing in it from a particular angle. Yes, he has lost the outside, the openness of out there; but he has also lost the inside, the in here.

Tutti

Of this, more presently. Let us first notice that his consciousness, of course, had never been one sense at a time. The field of his awareness was undivided by the distinctness of the portals through which his experience was channelled. There were synaesthetic cross-references – silent echoes in mirrors and invisible mirrorings in echoes, bright red scents and pungent scrawps – but we are talking about something more fundamental.

Consider a moment snatched from those many hours he spent on trains, commuting to the capital of the country of which he had once been a citizen. Vibrations in his buttocks, murmurs of the dialogue between the train and the rails, voices in the carriage and phones ringing, and the countryside flashing by and casting its shadows over his torso and papers, the taste of coffee in his mouth: these were all co-present in his presence to himself and of the world to him. The experiences were anchored in, and framed by, a body that exercised exquisite judgement in knowing how much to be present in its own right in order to tether his experiences to a solid here and now and yet not be so obtrusive as to occlude his access to the world around him. Perception was grounded in proprioception, underwriting a solid being at the heart of the 'where' in which he found himself, italicizing the inseparability of spirit and flesh.

The fabric of this solid being was woven out of quite homely material: abdominal fullness, pressure on his feet, dampness of sweat on his brow and both shoulders, adding up to the capital of a 'here' that makes everything else be 'there', and with some shoving, a little pushing, a little tugging, some effort, and some relaxation, giving added concreteness to his presence. There was a 'filling in' of the space between different bodily self-reports, so that the heaviness of an arm, the self-aware angle of his head bent over a page – the head itself deftly sketched in a few brush strokes such as an intermittent seepage of saliva, slight soreness in the eyes, a dialogue between a hand and the chin supporting it connected with a forearm pebble-dashed with goose pimples, and nostrils cooled by inhaled air – the faint strain of a back holding itself straight, trousers stretched over thighs tinting a patch of flesh with grey awareness, connected with aforementioned feet enjoying their warmth in shoe-softening socks. All of this found its place in their volatile, endlessly changing, sum, supporting the tautology of his first-person being.

Thus the miracle of a carnal self-awareness that somehow managed not to cancel or occlude the openness of the open to which it was addressed, with which RT had to engage. RT, as noted, had been explicitly surrounded: he had possessed his surroundings and been possessed by them. The opacity of the body in the bed belied the world once lit up within it or (less vulnerably) lit up because of it. His being in the world had been an in-this-body and out-of-this-body experience that the cold meat could no longer sustain. He had appropriated space and made it a world into which he was at once helplessly thrown, and its commanding centre, as its most salient item, or rather the presupposition of all salience. His had been a many-layered here – with a notional mid-point, a 'heremost', at the centre of a 'there' raggedly bordering elsewheres, composing places of work and play, awareness and oblivion, knowledge and ignorance, enchantment and frustration, concealment and revelation, privacy and publicity, future and past, possibility and actuality. This outside, the space of the ordered and disordered pell-mell of his life, was not the centre-less, aseptic space of geometry and of material objects – an emptiness filled with things like his corpse that obey the laws of motion – but a space of 'here' and 'there' and 'over there' and 'comfortingly' or 'threateningly' 'near to' and 'achingly' or 'thankfully' 'far away'. This space was the theatre of his days, of the million acts of the largely undramatic drama of his life.

And this, the condition of his finding or placing himself in *circumstances*, is where we must begin as we try to give the measure of the distance between RT and this corpse, between you, reader holding this book, and that worldless body in the mortuary ticketed with your name which will no longer remember you.

six

Space: Distance

The immensities of his world were the most literal manifestation of the distance between RT and the corpse that, courtesy of those who survived him, still bore his name. Vision had opened up his state to one of being surrounded. It was based on glances that revealed the openness as a network of distances. Those distances were italicized by sounds whose diminution to faintness expanded the scope of 'over there' to 'far away'. The visible limit to the visual field broadcast further distances, places out of sight, as did the silences into which sounds were swallowed.

There were many modes of near and far, of closeness and remoteness, of foreground and background, of here and there. Gathering all these distances into some kind of rational order – other than the sterile order of a purely quantitative vision that reduced places to decimal places attached to angles of the compass or axes superimposed on emptiness – would require an impossible ingenuity, and unachievable powers of recall. Scattered observations will have to stand in for an adequate reconstruction of even the most straightforward aspects of the outsides RT's corpse would mourn if it were not beyond mourning.

Let us stay for a while with vision, since it is the most obvious bearer of the richness and complexity and multiplicity of distances. The objects in its field were stitched together in a continuum of discontinuities, with the visible and the visibly invisible, the revealed and the manifestly hidden, being united in a manifold, populated by items with individual shapes, sizes, and locations. RT's every eyeful had encompassed many different kinds of things and places, each with their centres, peripheries, and layers. And eyefuls were, as we have noted, supplemented by earfuls, and anchored to a haptic universe built out of the interaction of his carnal being with the material world in response to basic needs, less-basic desires, and less-basic responsibilities – the serious and frivolous business of his life.

By this, most immediately, he had been located in and, more importantly had located himself in, space. He had experienced himself as next to things, on top of them, underneath them, a short or long way away from them, inside and outside them. Out of all these nears and fars, insides and outsides, he constructed the ever-changing immediate theatre of his life. Objects ready to hand, to foot, to mouth, or unready to hand, to foot, to mouth, the in and out of reach, were incorporated into the card-castle of the moment. Each item – table, window, tree – could in principle be allocated its separate place and, if he had drained the network of sense that held it together, various modes of here and there would be reduced to physical space and he and the items that surrounded him would be assigned locations in a coordinate system that in themselves had only a virtual mathematical relationship to each other – the kind of relationship (or non-relationship) his corpse has to the slab on which it is recumbent.

In the world of his daily life, where the remoteness of the far-off had not been reduced to numbers applied to struts in three-dimensional space, distant objects wore their distance and distant places were sometimes invested with enchantment. They belonged not merely to a point at the end of the virtual line of the gaze but were steeped in a spirit of place, in past and present meaning, were nodes in a network of significance. And, unimpoverished by the paint-stripper of the mathematical gaze, ordinary 'over there', viewed through a study window, would include a cat stalking a bird on a lawn, a person sleeping on a lounger, a garden shed pregnant with a multitude of artefacts, mossed roof-tiles, the accidental beauty of a blazing azalea next to apple blossom, a dance of shadows around the feet of a hedge separating addresses, neighbours, and family sagas, and a blown newspaper bearing a picture of an African village and an advertisement for the local eventide home where he might well have passed his last years.

An unimpoverished psychogeography of the space into which he had been cast would remember the interval between his sleepy childhood eyes and the sunlit tree-top leaves seen through the parted curtains of the bedroom window. Or the link established by a glance between an upland field where he had often stood on the way home from school, his hair tousled by a breeze, and a church spire dimming as the evening evolved nightwards. Or the interval between this church and another more distant, marking the middle ground and the background, drawing his gaze towards a tree-smudged horizon, suggesting further distances. (He had loved modest suburban rises commanding spacious but detailed views.) Or between the shadow of his index finger and the moon he is pointing at. Or between the back seat of a car smoothing down a highway and a plumpness of early summer trees, their green made more brilliant by a massif of black clouds, momentarily teasing him out of one thought into another remote from it. Or between a cloudy heaven and ploughed earth linked by a spoke of light. Or between his seat on a plane (as he looks up from his book) and the suburbs of the city he is visiting – where he will later be swallowed up into a quartier, a street, a building, a room, a conversation, a happy moment – flying backwards beneath him.

Since his earliest years he had been a lover of promontories, crags, high windows, tree-tops that made him a master of great spaces and all that was gathered into them. He had spent many hours of his childhood tasting afternoons from the shady vantage point of a huge, talkative tree, somewhere between a galleon and a cloud, rocking with the wind, feeling on the verge of flight, captain on the bridge in the high inland seas, looking down on roofs, on the pedestrians journeying through the streets, bald heads and feathered hats, on the tops of buses, or on fields where the breezes caused the light momentarily to silver tall grass stroked by the wind.

Distance was laid out possibility, the promise of the new and different, and, to varying degrees, of the exotic. Although in subsequent years he had travelled far beyond the childhood home, his strongest intimations of the enchantments of distance had been felt closest to his starting place. Sleeping in a room next door to his customary bed or exploring the uncultivated reaches at the end of the garden yielded the most acute sense of being 'far away'. And nothing could compare with those first bicycle rides as a ten-year-old. Self-propelled, travelling at three times walking speed down new roads, round unexplored corners, into adjacent suburbs, the penumbra of familiar bus routes, to parts of the city hitherto unseen, and beyond them, out to the countryside and to hitherto unvisited towns, gave him an intense sense of being 'remote from'. As he extended the arc of his adventuring, he *entered* distance, *became* far away.

His pleasure in remoteness took on a narrative form, a story of departure from and return to the home key. And distances opened on to more distances: the new road gives a view of houses on a hill; from those houses there is a view of hills beyond the city; and those hills command a view of the sea. He had a sense of the gradual unfolding of an ever more exotic world without limit. This was the erotics of distance; a pure sense of beauty as an opening up of the world to his gaze, as the assumption of a tor on being, on his days, on his life.

Nothing in his future travels could compare with this enchantment, because 'distant', 'far off', 'close' and 'handy' were matters of calibration. There were so many acuities of vision or envisioning: the town and the countryside surrounding it; a district within the town, defined by a pub and a church; streets dividing up the districts; the individual address within the streets; the house distinct from the garden, upstairs from downstairs, bedroom from bathroom; the chest of drawers from the double bed; the sock drawer from the shirt drawer; the left from the right sock; the inside from the outside of the sock. He had imagined once, sitting at a café table in Athens, the letter he was writing to his mother arriving at the right address, carried past the tree-planted and lawned and lamp-posted pavement, up the crazy paving path, into the lounge, and then into a drawer, where it would lie at the heart of the place he once known, awaiting his return.

There was no limit to the variety of the ways of granulating space – ranging from the gaps between stars to the space between the micro-organisms visible down the microscope that had occupied so many boyhood evenings. And being just out of range of Mummy's hand, or knowing that her kiss was downstairs, or the journey from home to first day at school, were associated with a greater sense of distance than any transatlantic flight could yield.

Our viewpoint – that of RT's corpse for whom all distances have collapsed – gives us permission, even something like a philosophical obligation, to remember or (if memory fails) to imagine the various ways that spatial intervals figured in his life. We may think of the interval between the itch on his nose, the hand that scratches it, the sound of footsteps downstairs, the eruption of a wood pigeon out of the flank of a tree he once dreamed of climbing, and the country whose civil war is being discussed, on a radio talking to an empty kitchen. Or imagine all the to-ing and fro-ing that added up to a fortnight's holiday, laid end to end, as the spatial correlative of the gap between the day of arrival and the day of departure. The links in the chain would encompass a visit to the toilet in the middle of the night and seeing the moonlit sea, a cliff walk, a thousand trickles of to-and-fro during an afternoon on the beach (ice cream, cricket, swimming), a visit to a country house. Or bring into focus a distant pub, visited many years ago, where in the lounge bar there was a lampshade whose little windows opened on to a coaching scene copied from a painting a century or more ago, of a real place some hundreds of miles away. Or connect up from end to end the journeys linking the second-form classroom in which he sat under Sir John Everett Millais's *The Boyhood of Raleigh* with, decades later, the moment of seeing the original of this picture for the first time in a gallery.

seven

Space: Partitions

The space through which he negotiated the long chain of journeys that had ended in this state where his openness had popped like a bubble was multiply divided. His world was portioned by partitions and barriers, and locks and keys, and vetoes, and permissions, and walls that parted, and barriers that melted, of exclusion and inclusion, of a million modes of outsides and insides, of public and private, of yes and no.

To step back a little. With partitions come new ways of extracting distances from space, already divided by woods, rivers, crags, seas. On to the crowded surface of the natural world are applied walls, hedges, fences, gates and doors (belonging to houses, cars, offices, sheds, shops, and pubs), screens, curtains – barriers to the body or to the wandering eye and the listening ear.

Space, being divided, was multiplied. Nooks and crannies served up by nature were augmented a thousandfold by bricks and edicts: by walls that divided field from field, farm from farm, house from house, room from room, garden from street; and by fences – the living veto of a hedge barbed with thorns or a dead foliage of wire sometimes spiked with metal barbs or electricity. These direct blocks to the wandering body were reinforced by warnings that unauthorized access punishable in law might lead to other constraints on his freedom.

Barriers discriminated between those who were allowed to pass and those who were excluded. They upheld property and propriety, defending what was owned, and 'my own', what was proper, against invasion. The Tom who peeped over a wall, or pressed his ear to it, invaded both. RT's right to wander over the surface of the revealed earth was thus closely regulated. Trespass was a primordial violation.

The dappling of keeping out and keeping in, of prohibitions and permissions was ever more complex in the towns and cities in which he lived. Writs ran but they also ran out, and curtilages were curtailed, though the limits were sometimes ill defined. While space was clearly personalized, personal spaces had borders of such complexity as to make the edge of a fog, or of sleep, or good taste, or of the semantic field of an abstract term, seem sharp by comparison. The script was sometimes barely legible but misreading might have calamitous consequences.

Walled, hedged, and fenced spaces had sub-spaces. Inside the house was the room accessed by a staircase, with a bannister rail supported by a file of uncomplaining caryatids, and a door. Inside the room was a wardrobe, which opened on to clothes. Inside the clothes were pockets. And inside the pocket was a sealed envelope that spoke its veto to all but one: 'Confidential: Addressee Only'.

The regulation of access to enclosed spaces became more elaborate and occult during his lifetime. Doors could be opened or shut but the possibility of the door being opened was itself regulated by keys, switching on and off mechanical blockage, maintained by the tenacious single tooth of a bar, a bolt, or the customized dentition of the wards of a key, themselves at the behest (in the later part of his life) of electronic signals which he knew how to trigger. His key ring had become ever more weightily hung with the means of accessing houses, rooms, offices, lockers, drawers, and cases, an archipelago of permitted access scattered in an ocean of spaces from which he was excluded. There were spaces to be negotiated with permissions; sometimes there were papers that had to be in order. The ever-changing contents of his key ring could have been the basis of a biography.

The doors against which hands knocked, the windows against which noses were pressed became ever more intricate and abstract. Some doors required pushing, others pulling; others were concertinas; yet others revolved, slid sideways or upwards, or rolled themselves up. Some required nothing of him, automatically melting before his bodily presence. More often, entrance qualifications, dress codes, memberships, were required to spring the locks defending insides from outsides. His wallet grew fat with swipe cards (key without skeletons) that convinced barriers that his business was legitimate. He acquired increasing quantities of security codes and PIN numbers to give the same reassurance when he attempted to enter physical or electronic spaces in order to act upon or interact with their contents.

The story gets more complicated as the dimensions of human space and the interactions of inside and outside become more tangled. Those who are inside, protected from the elements, plants, animals, and from fellow human beings, wish to have intelligence of the world beyond, to access the outside without going outside. Hence the perforations in the barriers – spy-holes, fanlights, opercula, oriels, windows.

The miracle of glass allowed cognitive access to the outside without giving the outside – bearing wind, rain, noise, beasts, and adversaries – access to the inside. Transparent partitions had been ubiquitous in his life: from bottles that revealed their contents to display cabinets (look but do not touch) and, of course, windows. Windows were modified in many different ways. Lightly veiled with a woven mist of net curtains, they allowed those who wished to look out to do so while frustrating those who, uninvited, wanted to look in. Frosted glass permitted light, but not gazes, to penetrate the barrier between within and without. Ripple glass likewise baffled any gaze seeking unlicensed intelligence. Windows could be cancelled in different ways: the blackout of curtains, Venetian-blinding with rows of lowered eyelids, shut-eyed with steel.

He had had many ways of regulating his visibility, from instructing another to close his eyes, hiding in a cupboard, round a corner, behind a hedge, running into the darkness, drawing down blinds, or leaving the country. And of increasing the visibility of the world from one of the million interiors he entered during his life: lifting blinds, cracking open a door, switching on lights, demisting the windows of his car (when he loved to see how the shrinking water tissue passed through a Venetian blind phase), consulting mirrors that allowed his gaze to bend round the corner of a road, peering through scopes of various sorts.

Learning to live at peace in the world required him to be sensitive to the complex boundaries of bailiwicks, and the wiggly lines between the private and the public, the outer and inner ramparts of apportioned spaces and the thresholds that could and could not be crossed, grass that had to be kept off, and zones that were roped off with visible and invisible cordons. He had been extraordinarily adept at negotiating the multidimensional, multiloculated open-and-closed spaces. Thresholds – cloisters, awnings that invited and awnings that forbade sheltering, ante-rooms, porches, reception spaces, common parts – were entirely intelligible to him.

RT was a master-navigator through partitioned space. He had thought nothing of entering his front door, rushing upstairs, passing through another door into a bedroom, opening a third door into a wardrobe, and extracting from the pocket of a jacket hanging in a wardrobe the car keys which would enable him to drive through a maze of streets to the house of another to whom he would deliver another key giving that other access to an office, a safe, a metal box, and hence to papers that were themselves a verbal key to yet another place in the world, a room and window that overlooked a landscape of carefully delineated possibility. And how adroitly he had maintained his poise in a multitude of interlocking and competing spaces, thinking (for example), as he drove to deliver the key just referred to of the radio play he was listening to, following the footsteps in the drama from one room to another in an Irish farm, as he turned left in order subsequently to turn right, in order to…And so on – ultimately, of course, to the place where he is lying now, where 'in order to' has come to an end.

Thus he had been surrounded, ensphered by space, and by spaces within space of such different scales. There was the great divide of earth from sky, the divisions of earth into 'heres' of varying size, defining and defined by 'theres' matched to them, in their turn defining by an horizon of sight and thought (vision and articulation being the greatest distinction in the partitioning of space) the 'Elsewhere' or 'elsewheres' that lay beyond. He had liked to think of the pie-charting of his world into a North that reached into the idea of snowy wastes, and unheard wind-blown pine trees articulating their desolation, a South of beaches and night cafés and life that spilt out into the warm streets (with so many barriers melted), or an East where the rising sun was a sense of possibility, and a West of elegiac tiredness, of conclusion, and evening darkening to night, the two 'ings' marking a slope he could not resist calling 'gentle'. And he relished the smaller partitions right down to the neat divisions in the filing cabinet where abstract themes, *arrondissments* of cognitive space, were allocated their proper place, or the pockets in his trousers where small change co-habited with fruit pastilles and the screwed-up remains of little notes to himself, messages to a future from what had become a past. Or the progressive coning down from the town, to the row of shops, the chemist, the back room where medication was dispensed, its cupboards and labelled drawers, boxes of boxes, bubble packs, and the individual pills that had postponed but not cancelled his appointment with extinction. The boundaries represented by sachets, purses, paper bags (often expressing more courtesy to the customer than the shop assistant), cupboards, suitcases, walled gardens, gated communities, postcodes, city walls, and national borders had been equally familiar to him.

The opposition between the outside and the inside was primordial and yet the outside penetrated to the heart of the inside. 'Outside' and 'inside' were as necessary to each other's construction as 'here' and 'there'. The partitioned space he walked through was as plicated and imbricated as the body that walked through it. Porosities were allowed, indeed necessary. The door knocker that said 'bang me' and doorbell that said 'press me' were available alike to family, friends, police, salesman, canvassers, and felons, so that newcomers could enter his life and new stories get started. There was ambivalence: the doormat that carried a standing 'Welcome' also implicitly requested that those who came from the outside should not bring the muck of outside with them, transfer the public highway to the private carpet. His attention was not infrequently on standby to be penetrated by the landline, by the buzz of a mobile in his pocket, and the influx of a boundless public space world through radio and television.

Which brings us to the most universal, powerful, and subtle generator of partitions, barriers, and walls: the focusing of attention. He passed through his life in a more or less permeable cognitive burqa that limited the seepage of the world into his consciousness. His vision tunnelled by preoccupations and prejudices, by attentional habits that were hard to notice, never mind to kick, he turned his trajectory through the open into a dimly lit passage connecting his first years and his last. And some blinkers were systematic, deliberately assumed, supporting an individually or collectively mandated aversion of gaze, a framing that prevented him at times from seeing what was in front of his nose. There was the unmentionable connected with the intimate and with certain parts of the body revealed only in certain situations described as 'intimate' concealed by barriers to the passing eye that a more prudish era mentioned as 'unmentionables'.

We have entered new, and problematic, territory.

eight

Space: Closeness

To be human is to be a becoming rather than a being; always to be on the way to being something else. In the light, or rather darkness, of the journey's end, arrival cannot be the point of all that has gone before. The later years are not therefore a privileged time realizing what in the earlier years had been only a potential. Childhood and adulthood are equidistant from completed sense. The narrative of growth, of increase, of maturation, of learning curves that point upwards, is offset by a counter-narrative of shrinking, decrease, decomposition, and learning curves that point downwards. No special value – that implicitly downgrades the remote past to mere *en route*, when RT was a preliminary, raw, unformed – attaches to the most recent years. Old age – dismissed perhaps as mere 'begoing' – here makes common cause with childhood. From the standpoint of his completed span, boyhood and adulthood were equally significant.

And so we are free to notice, and perhaps to mourn, that he had never fully realized that enchantment of distance he had sensed as a child rocking in the tree-tops above his childhood domain, or his boyhood joy in the hitherto unwalked streets of the suburb adjacent to his own when, his stride rendered frictionless by a bicycle whose wheels propelled him through space at unimaginable speed, he explored new worlds. In part, this was because flying and free-wheeling were always flanked by effort and trudge on the earth to which he always had to return. Busy making his way through space, distance was often a mere a barrier to what *had to be done*. The Eros of such distances, however, was all the more readily forgotten for being associated with the years customarily downgraded as the mere threshold of life, the years before he arrived at himself.

A joy that had no settled object, that was with the wind in the trees, the light in the window, or a gleam on a puddle, or the view from the top deck of a bus heading into the boundless crowd that is a city, into possibility, into a future that was still open, would always be fragile. Its moments did not add up to, or feed into, or belong to a story. And this was true, notwithstanding that few delights could match the feeling of being hidden – in dens, in woods, as he ran away from the one who played sentry calling up to a hundred (with eyes closed as the rules demanded), or crouched in the unintentional space beneath the stairs, curled up in a cupboard or chest of drawers, beneath the bed, under blankets.

Hardly surprising therefore that the enchantment of distance was appropriated early by another kind of enchantment. The delicious sense of possibility was narrowed into a more focussed longing, directed towards an increasingly obtrusive 'far away' in bodies that designated worlds out of reach, and placed him on the edge, excluded, even in exile. The erotics of distance were displaced by a frustrated longing for a certain kind of closeness, by the universal adolescent preoccupation. An often lovelorn, more often lustlorn, teenager, with a harem of faint hopes, he was a too familiar story. Eros appropriated the enchantment of distance and transformed

space. An entire city could become a vacuum, an absence, a solitude, a place of rejection, humiliation.

Carnal longing, the subject of a sniggering knowingness, of gossip and competitiveness, compressed the spirit of place into a small space; into a preoccupation with a figure, a shape, cloth warped by the pressure of warm flesh constrained by it. The vastness of the world, the magic of a thousand distances, were occluded by an inaccessible closeness intimated in the ellipse between the buttons of a blouse, by the black-stockinged darkness beneath a skirt, or an evening darkness in a voice or tanned skin bringing far places into a near room. Layers of here and there – house, room, clothes, guardians of difference and distance – were at once brought within, and placed out of, reach. How many partitions, oriels, walls, windows were summarized in garments that concealed and revealed a human body! How close-packed the contours of interiority seemed as he participated in the commonplace obsession with glimpses (caught or offered) of the gap between stocking top and thigh, of slopes of breasts, of creases in jeans that converged towards the anatomical epicenter of the objects of a free-floating priapism, of the (not always accidental) accidents of déshabillé that permitted the beach momentarily to slip into the street, the bedroom into the drawing room.

Coiled-up distances were intuited in another's warmth, the fantasy of a different world incarnate in desired flesh, as a lower lip summarizes all that the mouth might say. The living tent of the most familiar of all spaces – a territory defined in himself by an itch on the nose, a tear in the eye, a pressure on a stabilizing hand, the buzz of a toothbrush placing a mouth on a map, the heels pressing the carpet pressing back, a dappled fabric of warmths and coolths, folds, slopes, and dark places of the body – were transformed into an inapprehensible mystery he longed to grasp.

A new geometry of distance emerged, one defined by signs, by permissions and vetoes: the perilous ambiguities of physical intimacy located between affection and impersonal appetite, which could serve both pure love and commitment to a life more or less shared, and the ache merely to possess. The expression of desire, often undeclared, and if declared, clumsy, and risking mockery, had to be filtered through a lattice of logistics, courtesies and rituals. He could never be certain that the desire to touch and the desire to be touched coincided. He might be found wanting, unwanted, creepy, sinful, disgusting, even duplicitous. The prizes were doors unbarred and unbolted, zips unzipped, buttons unbuttoned, as granted permissions tipped the balance of concealment and revelation.

Now, as we examine his life from without and as a whole, granting all its phases equal standing, we are tempted to recast his emergence from boyhood as a loss, and the displacement of the unfocussed erotics of distance to sexual desire as a constriction. The mystery of the boundless out-there, of the land beyond the horizon signalled in the glow of the evening sky, or the sound of playing in the street heard from the edge of his childhood sleep, had been caged in the sense of a particular but unimaginable, and yet palpable world coiled up in the body of another, concealed under clothes and tantalizing him with implicit refusals. He might have wished to recover an Eros without ethical murk, without entanglements or a contingent prison of mutual obligation and expectation, that sometimes had little to do with joy, a delight far from the untidy place of bargaining, suffering, resentment, where adventure fizzles out into being-taken-for-granted in a world that is taken-for-granted, alleviated by deception, weighed down by boredom. The convergence of space on such particular places where so much significance was condensed might be construed as an augury of the ultimate collapse of space. A little death, perhaps.

Nevertheless, setting aside his share of the universal bad luck of ending, or 'ending up' as the phrase went, as a corpse after a finite life of incomplete meanings, he had been a lucky man. The arrival of Eros in his life did not eat up all distances, insisting that they should pass through the strait gate of carnal experience. He had known love and been loved in his turn; and the close-up immensities of a bodyscape may have been a cliché but it was a reality that had warmed his days. The loss of the childhood erotics of space and distance that had blossomed in the interstices of freedom in his anxious, busy, childhood had ultimately steered his life towards some of the largest and most precious facts that defined it: love, marriage, parenthood, and the translation of the unformatted torrent of experience into the lineaments of a biography.

And there had been moments when that forgotten delight known to his childhood had returned, unfettered by the helplessness of his early years: in gleams picked up from windows catching the last of a setting sun, viewed from the cheerful gloom of a tap room; in a panorama of river and beach and sea and church and sheep and roads and hedges available to the vantage point of a headland; on a hot starry night, in an interval of darkness between two villages located halfway up the massif of a mountain range, looking like seams of gold, where a flashing light prompted him to imagine the consciousness of the sleeping child within; in ploughlands bordered with bare hawthorn hedges scribbled on low dark and grey skies rifted with brilliance; or sensed, in the spring, spaces filled with birdsong blown a little by a May breeze that talked the million-tongued leaf-talk of the woods.

nine

Space as Theatre: Much Ado

RT had been only intermittently a tourist. His world had been for the most part a place not of contemplation and spectatorship but of action. It had been the theatre of that quantum of happenings that *he* made happen, in which he had deflected, or inflected, to a minute degree the unfolding of events in the material and social worlds in which he found (and sometimes lost) himself.

Now his passivity is absolute. He is not even lying down: for this body there is no lying and no down, though the thought of his being propped in the standing position – like the embalmed Jeremy Bentham in his sentry box – would be a mockery of the man who walked what we call the face of the earth.

All the ado by which he shaped his world, in dialogue with a world that shaped him, his winding path braided out of narratives unfolding through space and time, is now adone. This is not the place to pass judgement on his doings. We are not interested in his moral colour. He had addressed that, perhaps, in his declining years, enjoying mild satisfaction at his achievements and anguish at his failings, his misdeeds, his mistakes. He has joined the fallen who will not rise again and, as such, he is before us not as RT to account for himself but as an instance of *HS*.

Thinking of the dead as 'the fallen' – and their housing as a tomb etymologically connected with *tomber* 'to fall' – captures something fundamental to their impotence. We fall less as we grow up and fall more as we grow old; until we finally fall and stay fallen, outside space and time. We 'drop' dead and do not rise again. The living are the unfallen, taking their own path through life. Standing ('on your own two feet') is a metaphor of independence. It was the condition of his being able to act in the many different ways he did.

Standing is improbable – indeed it could itself stand for all the improbabilities upon which his life had been predicated – and had been achieved only by degrees.

In the beginning there was head control: that (now vacant) head learned to regulate its own position. Then there was trunk control. He was able to sit up and to take more, and wider, notice. Next he mounted to his knees and there followed in succession ever more effective modes of self-propulsion: creeping, crawling, snatching a step or two while retaining precarious balance, as if all paths were tightropes in the wind, then walking with increased steadiness, and toddling with some confidence. And so he achieved the archetype of the balance he sought in his life: tilting not to the left nor to the right, not forward nor backward. Ultimately, he was ready for full-blown striding out, running, jumping, leaping, standing on one foot while putting a shoe on to the other, nipping in, dashing out, marching (occasionally with 'bags of swank'), traipsing (a retrospective description of ambulation that proved to have been pointless, unrewarding, fruitless or just unnecessarily prolonged), dragging his heels, strolling, sauntering, straying, tearing off, and learning (even) the discipline (difficult for a new-minted being full of his sense of being alive) of keeping perfectly still. A few decades later, obedient to the implacable symmetry of the arc of life, these hard-won competencies of upright man were withdrawn. Trudging, staggering, limping, stumbling, tottering, doddering, shuffling, and crawling marked the downward trajectory. The staircase, an Alp to his toddling self, regained its mountainous aspect.

In summary, he got up and about, and was 'at large'; the large got larger; then he started to stumble; his deportment became time-bent and the large got smaller; and, finally, he was down and out.

The memories that correspond to those critical locomotor landmarks, when the blob in the cot became the self-propelled (and at times self-important) pedestrian, are lost, having been in the keeping of individuals who, themselves long dead, may be assumed to have no memories. None, at any rate, that the living can access. Equally lost to oblivion is the time when he discovered the joy of mobility for its own sake, expressed in tripping, skipping, hopping, and dancing, somersaulting, handstanding, hand-walking, accompanied at times by his own idea of music. Beyond recall, too, the time when he learned to subordinate his spontaneity to the multiple conformities of marching in step, mending his pace to others, working together as a team.

He had had an extensive repertoire of ways of standing: attentive, at ease, slack, slouching; head held proudly high or lowered in sadness, shame, despair, fear, or the pantomiming of any one of these; arms by the side, or akimbo (as if passing judgement on what he is observing, the limbs turned to handles on his torso), or folded (auto-hugging, sealing himself off from the world), with the pressure of his arms, each against the other, echoed in the mutual pressure of his compressed lips, or (most culpably relaxed of all) with hands in pockets, unready for action. Each of these postures had its appropriate sphere, were adjusted for the company he was in. To stand in one way rather than another was a signal, irrespective of its practical purpose. He could be ramrod when the occasion demanded.

He was a little late for the golden age of deportment – head up, bags of aforementioned swank – but he would have had no difficulty diagnosing aggressive, deferential, resentful, even grovelling modes of standing in others.

He was a connoisseur, likewise, of modes of sitting. Every chair is now vacant of his presence, or the possibility of it. For most of his seventy-five years, he was a fluent sitter: from upright, courteous and attentive, with lightly plaited legs, to slouching, weltering, sprawling, and even the often criticized *rocking back*. For sitting, too, was a signal, and was subject to a set of rules that were at least as complex as those that dictated when it was proper to stand, or to stand back, and what distance to stand at. And sitting was of course a necessity, because his body was not just the agent of his will but also a material object that placed physical pressures on the subject it embodied, not dissimilar to those that faced him in the obstinate material world.

Upright man, that is to say, can stand only so much standing and eventually craves the invitation 'to take the weight off your feet'. Relieving his lower limbs of the burden of supporting the rest of his body was one of the many free pleasures his life had offered him. He had childhood memories of houses where the lounge (an amusing transferred epithet) was called both 'the living room' and 'the sitting room', to acknowledge the extent to which living and sitting overlapped. Not for nothing was the seat the item most closely associated with government, and social power imagined as being exercised in the sitting position ('the seat of power') and those parliamentarians who wished to be independent of party whips were referred to as 'cross-benchers'. Sitting was not always so dignified. He had sat many times, waiting to be delivered of the contents of his colon, both agent and patient, spectator and participant, beneficiary and prime mover of the daily ritual on the hollow throne.

He had appreciated the legislated kindness of street furniture anticipating the aching legs of citizens. Long speeches and nowhere to sit down were markers of totalitarian regimes. His walks in parks were punctuated by benches commemorating the dead, who thereby dispensed in their afterlives the comfort and convenience of a good sit down. And some mountains, where he walked for the sake of walking and in order that he might see for the sake of seeing, offered weatherproof seats 'for those with bellows to mend', though here he was often content, as he ate his sandwiches, and allowed his consciousness to dilate into the great silences cupped between fell-sides, to utilize the immemorial bottom-ware his ancestors had fashioned out of depressions in the grass, groined rocks, and smooth tree stumps. Ever the opportunist, he requisitioned walls, stairs, kerbs, and branches to park his posterior. With less dignity to maintain, he would as a child often be found squatting (sometimes on the front doorstep of his house, awaiting someone's return), or cross-legged on the floor in rows of peers with arms folded, his upright back and attentive head broadcasting conformity to the minutiae of Respect for Others, or perched in a tree, or on the shoulder of a parent. The earliest seat, close to where he had entered the world, was his mother's lap.

The joy of sitting in an ergonomically friendly receptacle – possible even when he was journeying, courtesy of appropriately furnished beasts of burden, wheels, sails, and engines – was even more prized than the luxury of stretching, of crossing one leg over another, of twiddling his toes, of supporting his head in his hands, of taking a deep, delicious breath, of turning to a new position, or scratching an itch, or of bathing his eyes in a little dose of darkness: these were some of the more innocent ways of enjoying his carnal self-presence.

Unlike *HS* for most of its history, RT had lived a greater part of his life folded into the sitting position than in standing, walking and running away, toward, or about. With successive generations, more of the labour given to the world in exchange for the means to live had become sedentary. And so he passed much time in chairs, with or without desks and tables, listening, talking, learning, teaching, reading and writing. From his earliest years, the parked bottom had been recognized as a condition for concentration on a lesson or a story: 'Are you sitting comfortably? Then I will begin.' As he aged, sitting became increasingly necessary. Unhappy legs or general tiredness translated into thirst for a place to sit down and the hunt for such was as purposeful as for food and drink. He was grateful when seats in public transport were offered to *les mutilés du temps* or a stranger, quantifying his antiquity at a glance, gave up his seat. The more obtrusive operation of Newton's Third Law of Motion, according to which action and reaction were equal and opposite, required him to be more assiduous in shifting his weight from buttock to buttock and palliating the mutual pressure of body and support by a literal cushioning that spread it more widely over his hind quarters, in order to avoid asphyxiating his skin and consequent pressure sores. Lounging may have been a strangely passive action but it was important that he did not give way entirely to passivity.

The rules of sitting, like everything else in his life, were elaborate. Who gives whom permission to sit – 'Take a pew', 'Don't stand on ceremony, dear chap' – and the circumstances under which he or another might remain seated were key elements in the choreography of the togethering that was his life. He had early internalized the commands to stand up when a senior or a lady came into his presence so that he did this without thinking. Standing stood for so many things, gathered up in the notion of 'showing respect' – irrespective of whether respect was felt. He was of the first generation that found it easier *not* to stand up when the National Anthem was played – linking two gigantic abstractions, My Country and Love targetting the person of the sovereign. His remaining seated was, of course, a sign that he stood up for other things: for democracy, for his rights, and the rights of others, and against illegitimate hierarchies, the traces of feudalism that propelled certain individuals of severely limited capacities into ceremonial positions of authority on the basis of accidents of birth.

He had been present at a memorable concert in Stoke-on-Trent in 1976. The first piece – Rossini's *Thieving Magpie* overture – began with a drum roll, similar to that which used to precede the National Anthem. Respectful Potteries knees were jerked, thighs were aligned with calves, legs were extended, and soon most of the audience was upright in a gesture of a kind of loyalty that, in these densely provincial parts, had not yet been entirely phased out. The incident rippled through the next decades of his life as he referred to it, perhaps unfairly, to illustrate something (blind loyalty, subservience, automatisms) of the citizens of that unfashionable town, the idea of which was so far from the idea of London, though when one got down to details – a cobbled street whiskered with weeds, a queue for a coffee, birth, copulation, and death – differences were less apparent.

Notwithstanding the importance of sitting in his life and the complex dances that sometimes took place around chairs – insisting that he or she have a seat, drawing up one for a third party, being invited to join this table or that, preparing for a meeting by setting out chairs – much of human ado, in work or play, presupposed, as already noted, what it has always required since *Homo* became *erectus*: standing and walking.

Standing still was the condition of one told to wait. Waiting – the most basic of the 'much adon'ts' that had constrained and hollowed out his life of freedom and action – was pervasive. Keeping still was sometimes unbearable and his impatience spilt over into many grades of fidgeting that could formalize itself as 'pacing up and down'. And sometimes, by contrast, he had stood not merely still but stock-still, wanting to observe, hearken to, and take in, the world without the turbulence of his own movements curdling his consciousness or alerting its objects. The world seeped through the sides of the stock-still as when he performed his daily ritual with eye drops. It was under such circumstances that he arrived at a sense of being truly at a Mediterranean town where he was holidaying. Upright, eyes closed, pinching the bridge of his nose to prevent the drops being washed away by tears, feeling the pressure under his bare feet, the phial in his hand, he could hear with special clarity, through an open window, the intermittent sound of traffic, the sea, the multitudinous cries of children playing on the beach, and behind him the sound of breakfast being prepared. He was in the South of France.

'And did those feet?' Most certainly they did. Looking at those feet, now poking sockless and shoeless from beneath the sheet, is an invitation to retrace the steps they have taken and the different places they have taken him to. Thinking of what his feet did before they, as it were, hung themselves up, takes us into the progressions, regressions, and digressions of his brief trajectory between assuming the upright position and joining the fallen.

The temptation to slip into pedometry is strong and, given that the only way to deal with temptation is, as Oscar Wilde said, to succumb to it, let's satisfy ourselves with a couple of calculations concerning RT's ultra-marathon between the onset of toddling and the end of doddering. It is a conservative estimate that in the course of an average day – pottering around the house, wandering into the garden, shopping, executing the pedestrian ends of his journeys to work, moving between one hospital building and another, pacing up and down in his office, hurrying to clinics or meetings, nipping out to the loo, and doing ward rounds – he might have covered at least five miles. In seventy-five years he would have clocked up about 150,000 miles or six circumnavigations of the planet of which he was such a minute part. This total could be achieved without set-piece ambulations, walkings for walks' sake, such as strolls in the park, runs on the beach, rambles in the countryside, or miles covered, and doubtless counted, in walking holidays. He would have commuted between the upper and lower storeys of his house on average at least ten times a day – more on the days he was at home, less on the days when he was at work. By this means, he had completed a quarter of a million ascents and an equal number of descents, amounting to 1.25 million metres or the equivalent of over 150 Everest conquests.

We note in passing that his feet did other things, such as kicking footballs (with less than the median level of enthusiasm), bicycling (passionately), climbing trees (addictively as child), digging (reluctantly), and kicking living creatures (very rarely and then, with no exceptions, entirely playfully). On at least one occasion, he had subjected a stone to a *coup de pied* in a rehearsal of Dr Johnson's point-missing refutation of Bishop Berkeley's idealistic claim that insentient objects exist only insofar as they are perceived.

Such are the exotic uses to which humans put their bodies. But this is to distract from the larger theme of journeying though the multidimensional spaces he and his conspecifics had landed in and constructed; from the different ways he trotted the globe, leaving minute scuff marks on the face of the earth, prints in a muddy road in Northern Nigeria, crumbs of mud in a friend's home, wet marks on his own bathroom floor; and the different reasons that directed his perambulations.

In propelling himself on his hind legs, RT had joined an immemorial tradition that had begun long before man's most recent manifestation *Homo sapiens* put one foot in front of another. Humans have walked, run, stumbled, hopped, skipped, trudged, staggered, squelched, tottered, galloped from place to place for millions of years, since *Homo* became *erectus* trying by honest footwork to link hope and its objects, in pursuit of things to answer the biological imperatives or, in recent millennia, other imperatives that have arisen out of their infinite, collective capacity to transform the biological givens into needs and desires not envisaged in or even endorsed in nature.

And so those feet most certainly did. But *where* did those feet? RT had walked over linoleum and rugs, over soil and flagstones, over tarmac and laterite, lawns and glades, rocks and sand, on stepping stones and seabeds, over fell-sides and crags, over the earth with its grades of rumpledness, scales and registers encompassing bumps in carpets to mountain ranges. The bedroom rug, the pavement outside his house, the parquet floor in the schoolroom, stone-paved cloisters, flower beds and shrubberies where he searched for lost cricket balls, meadows (in order to be up to his calves in buttercups), sand dunes and rock faces for the love of it, woodland paths, screes, tracked and trackless wildernesses, swaying wood bridges, decking – they all gave his footsteps a different feel, and afforded the body propelling them a variable sense of itself, and made his gait speak in different voices. Barefoot, in stocking feet, slippered, loafered, clodhoppered, in shoes so smartly polished he could imagine his distorted face mirrored in toecaps, rubbered in Wellingtons, mountain-booted – there were so many ways in which he could interact with, taste with his feet, and the ground beneath them, as it presented itself in the rooms, houses, streets, meadows, fells, seafloors, into which he took himself, in stations, airports, piers, buses, trains, planes, ships, in suburbs, villages, cities, woods and meadows, countries, continents.

While to the ears of familiars his gait sometimes had a familiar cadence, the accent of his feet would vary from surface to surface: the whispering of slipper on carpet, the clipped authority of solitary steps as steel-tipped shoe met granite pavement such that his noct-ambulation was broadcast into the silence of a deserted suburban road (often accompanied by a long noodle of chords as he whistled to amplify his presence to himself), the sucking soggy sound of his booted foot extracting itself from the glutinous black fudge between tussocks on bogland. And then there was the pleasure of crushing downtrodden, fresh-fallen, frozen snow, permitting an ideal crispness to find its voice, where the word itself has something of the sorbet in it, such that a spoon could enter the 'crisp' speaking 'ness'.

Snow-walking had been most perfect example of ambulation-as-footprinting, where every step leaves a three-dimensional negative of the shod foot that had made it. His prints – those mass-produced, unlimited edition templates – were almost as transient as his passage through the place where he had left them. Nature would efface them, as the snow he had violated would melt, rain would shape the mud to another shapelessness, or the tide come in. And so would the steps of those who followed him, overlaying his traces. The signal of his track was merged with others' and consequently tousled to something lost in noise. (Thus do descendants wipe ancestors.) Eventually prints make paths whose firmness resists registration of individual path-takers. A path created by past journeys would guide and ease those to come: the most emblematic of the gifts he had owed to his ancestors, recording their collective presence and effacing their individual tracks.

The surfaces on which he had left his footprints had in turn left their imprint on his feet. He was a horny-footed son of ambulatory toil, his plates pounding the earth and being pounded in turn, so that those heels, on the few occasions when he scratched them to satisfy a rare itch, rasped. The percussion of each step had thickened layers of *stratum corneum*, a patchy auto-shoeing, by the farrier of his own past, arming his foot against sharp stones, thorns, and the mandibles of crushed insects avenging their coming death with little acts of spite. But the minutest inspection of his heels would not have betrayed what he had walked on or through: for example, a mountain track high above the sea in a quiet Greek island; even less his joy at its pine-scented silence. Neither would there be any echo of a certain walk through the city at night and the sadness and fear that he felt. There was no plantar record of his irritation as he ran up the stairs for something he had left behind in his rush to get to work or the anxiety that had accompanied so many of his journeys along hospital corridors during his forty years as a would-be healer.

The muteness of his heels was a reminder of how little could be inferred of his life from his corpse or, come to that, of the life from the living body or (except in the most general terms) where and why the weather had beaten the weather-beaten face, or of the thoughts behind that face as it had confronted the rain of glances emitted from the faces of others. Those feet had shown their long drawn-out past a clean pair of heels.

His hithering and thithering was on many scales – shifting from foot to foot, pacing up and down, up the stairs, out to the shops, down to London, in and out of the country. Such was the ubiquity of travel in his life, it is hardly surprising that the misleading cliché of life as a journey – linked to the idea of it as a quasi-linear story connecting the howling amniotic-fluid-drenched entry to a scarcely less dignified exit – has such currency.

We can dismiss the myth of linearity right away, even in the case of literal journeys from A to B. They bore little resemblance to the geometrical line AB. Travellers are never crow-fliers or bee-line. Taking the bus into town requires walking in the wrong direction to the bus stop where the bus going (roughly) the right direction is to be caught. Once on board he would walk against the overall vector of travel to find a vacant seat, engaging in the elaborate choreography of squeezing past others. The bus itself meanders in order to hoover up as many people as possible, taking a roundabout route and pulling in and pulling out of stops, circumnavigating other traffic. Finally, RT would walk backwards from the nearest stop (which happened to be past his destination) to reach point B, a rather blurred target – a street, an institution, a building, a room, within which there would be much to-ing and fro-ing, before eventually he arrived and sat down.

And as for those air flights he so often took in pursuit of academic glory – where, for example, point A was the desk in his English study and point B the podium in a lecture hall in Singapore – it is quite sufficient to note (in order to develop our case) certain components of the journey: the rush back upstairs to fetch the forgotten passport; the circuitous processes of checking in, going through security, passport control, immigration procedures; the passacaglia as he squeezed past other passengers bent on other destinations, all the complex choreographies of negotiating other pedestrians on foreign pavements; walking down a corridor to find the relevant lecture theatre, taking the few steps up to the stage. And none of these items had corresponded singly, never mind when summed together, to anything that could be portrayed by a straight line whose stages were aligned with the great geodesic of his pilgrimage from A to B. Journeys were to some degree straightened out only by being recounted, so that a continuous track seemed to be traced across a hypersurface of meaning, action, and living. But even this simplified story lacked rectilinearity on account of the obligation of the reader's gaze to scan back and forth, or the cocked ear to listen, retain, and recall, so that braid of narrated sense linking a temporal start to a temporal finish might be woven.

What's more, his journeys had been bristling with secondary tasks, fitted in 'to make best use of the time', particularly when he was passive – being propelled or waiting, not otherwise engaged. Each digression – getting a coffee to improve concentration on the academic article he was reading or writing, making phone calls in a quiet place, going to the toilet to bank some space in his bladder prior to the Big Talk, buying a paper to catch up on the news – had such wiggles that a line drawing of these incidental trajectories would take a lifetime to replicate. And these did not include opportunistic activities – enjoying the view, engaging in conversations with fellow travellers, reading a book – while his intentions free-wheeled towards their fulfilment. The visible digressions tying knots in the trajectory of any journey were supplemented with invisible inner ones, in which memories and thoughts and feelings – solicited and unbidden – decorated his geodesic with absences that ate it away from within.

So the journeys undertaken by RT had had wiggles, and wiggles within wiggles, and wiggles within the wiggles within wiggles. Some wiggles had been involuntary: imposed diversions or the consequence of navigational failure. Because the theatre of his agency was so much bigger than his body and its sensory field, it had been not unusual for him to get 'hopelessly' (the customary exaggeration, for hope never deserted him so long as irritation was alive) lost. Instructions, maps, guidebooks, even compasses, proved insufficient to connect him with destinations where tasks could be performed, goals achieved, and duties discharged. Under such cir-cumstances, the opacity of the material world seemed to be in direct opposition to the will that informed his purposive walking. Even so, he still arrived at most intended destinations.

As he was a child of the twentieth century, his journeying was not just a matter of walking, running, and stumbling. Nor did he rely on animals to transport him from Point A at time t1 to point B where he was required to be at time t2. If he had ever ridden an equine, it would have been a donkey at a fair or on a foreign holiday. For he was the beneficiary of the invention of the wheel and, more impor-tantly, the axle, which had liberated legs from the burden of travel. His luck was to have been born late enough for bio-powered trans-port – drawn by beasts or by humans treated as beasts – to have been replaced by beastless vehicles that had found an equivalent of their vital forces in insentient engines harnessing mindless energy to serve the aims he had in mind. Horses had been replaced by inorganic proxies delivering power, traction, and propulsion; four legs by four stroke; grass by petrol; and the stench of dung by that of exhaust fumes. He could therefore travel while remaining folded in the seated position – indeed securing a seat was often the most effortful portion of a journey. Such walking as was required of him was therefore usually restricted to the beginning and end of many journeys, the fine tuning, the bespoke particulars, that connected generic routes (the 22 tram, the intercity train, the intercontinental flight) with very specific destinations: the rooms where he was to perform, meet, eat and sleep. Journeying consequently became a matter of timetables, of booking, and of catching and not missing. Hands, to wave buses down, to leaf through schedules of trains and planes, to purchase and proffer tickets, to trigger the keystrokes that delivered bookings electronically, and mouths, to communicate needs and establish itineraries, became as important to travel as feet.

In the final years of his life, a different kind of travel started to predominate: virtual travel via keyboards contacting data stores, permitting electronic tours of an earth Google-mapped and home delivered to his sedentary gaze, the word and the finger superseding the feet, and of course permitting him to speak to remote interlocutors. The world he had left was speeding into a frenzy of downloads and uploads, not knowing where it would ultimately lead.

The ease of travel illustrated the extent to which pushing, pulling, lifting, dragging, various degrees of ambulatory hastening, had become less apparent than hitherto in his own life and indeed that of his fellow citizens. The agony in his arms as, taxi-less, he had carried his suitcases from the station to his student digs was a distant memory as, later in life, he more usually disembarked from a taxi and wheeled a wheeled suitcase up a short drive. Work had become, increasingly, a matter of driving pen over paper, or key-tapping, than stone-breaking, digging and chopping, banging into shape, lifting and lugging or chasing.

The sedentary life carried dangers, broadcast in the engrossment of the bodies of contemporary humanity weighed down by themselves but perhaps more importantly expressed in the invisible silting of vital channels delivering blood-borne necessities to his tissues. Spared the immemorial curse of having to live by the sweat of his brow, a lotos eater by the standards of the penal colony of the world and of history, he had undertaken, on a regular basis, activity that had brow-sweat as a primary aim, in the hope of increasing the total of his days. He would be found from time to time on a treadmill, running precisely and consciously nowhere, flanked by other glistening elders with equal seriousness and firmness of purpose, in order to make his heart beat faster, to rev up breathing into panting, and cause legs that no longer ached in pursuit of particular purposes, to ache purposelessly in pursuit of their own fitness for unspecified future purposes – or for the possibility of continuing to have purposes. 'Weights' were bought or borrowed in order to instantiate the essence of the burdensome, replicating the primordial burden of biological existence lived out in a gravitational field.

While RT had often undertaken journeys for the sake of looking and seeing, peering, gawping, spectating, he had travelled for the most part in order to arrive to perform some act. Any attempt to impose order on, or to create a *catalogue raisonné* of what he had done before he came to his present (utterly passive) pass, would be hopeless. Some general comments, supported by (frankly) random choices will have to suffice. This said, there is an obvious first port of call in our celebration of his agency: those hands, now pale as candle wax at their fingertips, unhaunted by intention, fixed in a posture that is neither a gesture, nor a grip. They had been the most significant agents of his agency, to the extent that their internal hierarchy of dominance and non-dominance, *major* and *minor,* senior and junior, had conferred upon his world a pervasive asymmetry. It was not for nothing that the role of his hands – working separately and together – had been acknowledged in the word 'manipulation'. So much of his life involved this 'tool of tools' as Aristotle called it: *HS* was, above all, a manipulative animal.

The erstwhile versatility of these ceaselessly gripping, grasping, grubbing, groping, holding, cupping, catching, pressing, pulling, pushing, squeezing, pinching, tweaking, twiddling, plucking, prising, picking at and picking up, carefully or idly fingering, leafing, tousling, dabbing, caressing, scratching, slapping, punching, embracing, stroking, patting, smoothing, screwing, tapping, shaking, drumming, clapping, poking, prodding, pointing (at, out, to), enumerating, waving, threading, scraping, insulting, bending, twisting and stretching, glad-handing, hand-shaking hands in shaping, exploring, and communicating with his world can be only gestured towards. To describe just the varieties of reaching, or even reaching down, would have been beyond the range of his mnestic and descriptive powers. The sheer number of modes of even minor manual actions such as waving, fossicking, and clapping would have defeated any attempt he might have made to list them. Ironical and melancholy waving, fossicking in pockets, rucksacks, or any number and kind of dark spaces, clapping enthusiastically or sarcastically – had all made their small demands on his time.

The fluency with which he combined successive grips in the most ordinary acts such as threading a needle, tying his shoelaces, or cleaning a cupboard, under the non-canonical circumstances of everyday life, would have defeated a robot regulated by the most densely woven fabric of ifs and thens. The subtle cooperation of his two hands in the screwing and unscrewing of caps on bottles, with the power-gripping left hand holding the bottle steady and the precision-gripping right hand doing the subtle business, was exemplary. The exquisite teamwork of his fingers causing a pen to dance out abstract meanings in filaments unwound from drops of ink, was no less extraordinary for being the result of much practice. The years of penmanship had left their own memorial plaque: a callus on the middle finger of his right hand, echoing the aforementioned heel thickening caused by literal as opposed to merely verbal journeying. And the hand had been complicit in the rather Byzantine relationships his body had had to itself. He would (to take a random example) deploy his tongue to lick the hand to transfer saliva to the pads of his own fingers with the aim of altering their coefficient of friction in order to improve their ability to separate thin sheets of paper.

In common with his pre-machine-age ancestors, RT had increased the power and precision of his hands through practice that enhanced their brute strength or made them more intelligent. But he had countless other ways of magnifying his manual efficacy, beyond the immediate assistance of kit such as tweezers, tongs, hammers, knives, axes, screwdrivers, spades, scissors, and secateurs. Minor amplifications included smiting objects to make them shout, as when he awoke an entire household through banging on a door with a knocker; and greater ones including tapping keyboards to send his thoughts beyond the reach of his larynx, and turning keys or switching switches or pressing buttons to activate some mechanism which would deliver heat, warmth, and other utilities and desiderata to the world around him. (So much was possible at the press of a button, so much in his life had been mediated by the flick of a switch, that he and his contemporaries could reasonably have been described as Button-Pressing, Switch-Flicking Animals.)

Given their centrality to his agency, it is hardly surprising that he looked after his hands with some care. He chaperoned them during their engagement with the brutish stuff of the material world. There were gloves to protect them against cold (mittens by which they harvested their own heat, occasionally supplemented by his hot breath), heat (oven-gloves), aggressive vegetation (gardening gloves), the rough surfaces of bricks or stones that left his fingers not liking the feel of each other (work gloves), water, corrosive substances, or contamination with vectors of illness (rubber and plastic gloves). By these means his hands had mitigated the grasp of the material world reciprocating his own grasp.

Gloves had not played as big a part in his life as they had in some of his fellow grippers. He had never owned boxing, cricketing and other gloves that modified the brutality of sports whose primary aim was supposed to be victory over, rather than damage to, opponents. Nor had donning these manual leotards served the secondary purpose of signalling in a small way the kind of person he was or his sensibility to the kind of occasion was attending. He had been aware of course that dress codes included glove codes; that evening gloves and gauntlets allowed the hand to speak in profoundly different accents; and that there had been a time when to have driven without gloves would have been a solecism. Gloves off, the hand was rough and ready, and ready to be rough, though this had definitely not been RT's style.

His hand had had powers additional to those mediated through actual contact with the material world. He had waved farewell, wagged his index finger, or snapped finger against thumb. These gestures, like many others, were richly polysemous. So much could be communicated through the manner in which he waved someone off – casually, enthusiastically, even (as already noted) ironically. Finger-wagging (which he had used sparingly, aware how irritating it was) was an all-purpose highlighter of what was being said, raising the font of utterances (bold, capital upper case), as well as being a stand-alone warning or expression of disapproval. Finger-snapping could: tap out a rhythm; support his teenage self-image as a pop star singing to an imaginary audience of adoring teenage girls; illustrate the speed with which something happened 'just like that', 'in a trice'; and could have summoned a waiter or someone obliged to wait on him though he had refrained from such discourtesy as an abuse of what powers he had.

Pointing – to a shared world – had awoken the possibility of another theatre of contact, beyond the body and the natural world, beyond arms to be scratched, rocks to be grasped and fellow humans to be gesticulated to. This is was the transcendent, invisible world to whose far reaches prayers had been directed, sacred spaces as absent from this corpse as from the world in which he had lived his daily life. For some of those mourning him, believers in the afterlife, and in an eternal RT, he was now in that world, having shed a body regarded as mere soul-rind.

There had been important interactions between these now inert hands and this inert body. His hands had supported his head when he was tired or weighed down with despair, had washed his face, had cleaned his teeth, and had combed his hair. His hands had concealed the contents of his oral cavity as he spoke with his mouth full. They had wiped his tears and blown his nose and reamed his ears. They had given him pleasure and relief as he satisfied (and cancelled) itches between his toes or under his armpits – creating a relatively rare internal carnal dialogue and one that he might exaggerate to mock those simian fellow-humans rugby players – or in his crotch, on his legs, on his arms, and by this means highlighted so many territories of his body to himself.

The appropriateness or otherwise of his scratching those territories covertly or visibly underlined the differentiation of his body into public and private parts. A particularly special itch had its capital in one of those parts – the most private, the last to be uncovered to a stranger's gaze, though such concerns with concealment no longer exercised him. Awakening and satisfying (and cancelling) this particular itch had requisitioned thoughts, memories, images, swathes of his relationship to the social world of which he was a part, imported into the loop between the hand that grasps and the body that is grasped, a circle closed by the identity of the one that acts with the one that is acted on.

The creativity of that now still hand had been enhanced by the kind of self-consciousness that had also enabled him to clutch hand to forehead and, by mocking melodramatic gestures on and off stage, to capture a fold of the collective consciousness of which he had been an inlet. He had been able to play with the interaction between manual gestures and language, by scratching his head as a parody of the standard metaphor of puzzlement or of hard thought aimed to resolve puzzlement. His thumbs, now forever untwiddling, had occasionally cooperated in enacting this topological mystery in which each encircled the other – to illustrate inner emptiness, impatience, lack of office, role or purpose, or indeed, the constraint of being kept waiting. Those hands had had a wide repertoire of duets – aping surrender, or scare quotes (in turn to signify or parody modes of behaving) or prayer. The wandering hands returned to their home base when they clasped each other to generate and exchange warmth, for the sheer pleasure of contact, or for making his glee haptic, visible, and audible (louder when dry) to itself and others when he rubbed them together.

In a world dominated by words and machinery, his hands had still retained their supremacy as the indispensable agents of his agency, the final common pathway of much of what he had accomplished. But while the person RT had remained in close touch with the organism RT, through the courtesy of his hands he had lived a life far from the biologically prescribed existence of other organisms.

Whatever he got up to, his hand had usually had a hand in it. Until, that is, he had been forced to hand over to others. The baton of agency was now in others' hands.

As with journeying, and with manipulation, so many of the things RT did were done in order to carry out other things, if only (as in the case of tourism) to see others doing things, at work and at play. Most basically, he moved his body – whether from Manchester to Singapore or from downstairs to upstairs – to *position* himself in order to do or to observe. Even the most porous schedule still had an aim: the encounter with the unscheduled, happening upon this and that, in the spirit of *placet experiri*. But most of RT's life had consisted of more or less focussed 'in-order-to-doing'.

His days had been dense networks of intermediate goals. He earned money in order, among things, to be able to afford to live. He did x or y or z in order to earn money. He underwent training in order to maintain his competence at x or y or z. He filled in forms in order to register as a trainee. He made a phone call to request that the form should be delivered to his house. He walked downstairs and asked people to speak a little more quietly so that he could make that phone call.

The unpacking of this 'in-order-to' is endless, of course, because the various parts of his life were not sealed off against one another. Everything had been a syncytium of connected deeds, rooted in a humus of presupposed meanings derived from a multitude of linked intentions. Sometimes the journey to the goal had seemed to presuppose what the goal, if achieved, would bring. For example, he had needed to eat in order to do those things that enabled him to earn enough money to be able to put food on the table. This circle was not vicious, however: others had helped him to 'stand on his own two feet'. He had begun his life as the unreciprocating recipient of care from those who had given him his many different starts in life. He had begun with a starter pack and his mother and father had been the prime bestowers of that starter pack.

Just how long teasing out the strands of the 'in order to' would take would also depend on how fine-grained the inspection and description of his actions was to be. But a careful look at any action – cooking a meal, tidying a room, going shopping, completing a clinic – would yield a long sequence of steps, each composed of a multitude of distinct movements. Take that phone call to order a registration form for the training course. We can think of his picking up the phone, dialling the number, taking a deep breath, articulating the sentences that count as the request, listening to what is being said, taking down the name of the person spoken to (in case something goes wrong), courteously thanking him or her, and putting the phone down, as distinct actions. And each of these would draw on larger actions or activities: namely the acquisition of skills and various modes of understanding. Learning the right kind of phone manners, say, and of how to make an articulate statement to a disembodied voice whose prior knowledge of your concerns is uncertain, and grasping the concepts of telecommunication, of registration forms, of courtesy, were all presupposed.

Even 'closed loop' actions, such as RT's laundering himself at the beginning of the day, involved many steps and entrained numerous elements. Compare the unmediated self-attention of a cat grooming its private parts with RT washing his hands (a strange wrestling match, a mock civil war between enantiomers, with no clear victor when the towel is thrown in) using soap obtained at a (relative) convenience store a couple of miles and many hundreds of deliberate movements (donning and buttoning outer garments, backing the car out of the drive, numerous gear changes) away. Purposes dissipated down networks of 'in order to' lost their savour and were shadowed by the ultimate disorder that lies at the end of all 'in order to', at the ocean of purposeless happening into which his trickle of agency would finally dissolve.

Actions fitted into bigger actions and they fitted into bigger actions and into bigger actions still. Forethought, illuminated by hind-thought, was materialized into the carnal web of his agency. The images of Russian dolls, or of fractals with patterns being replicated as the gaze drilled down to ever more fine detail, were not entirely off the mark. Or the computer engineers' notion of 'self-embedded sub-routines' with a *mise en abyme* of smaller and smaller components replicating the structures of higher-order actions. There was a melancholy echo of the hierarchy of subordinate and super-ordinate parts also seen in the anatomy of his now defunct body, in its individual organs, in the cells of which they were composed, and the organelles within the cells.

Even the life of letters – seemingly least compromised by the constraints of the material world – had broken down into an hierar-chically arranged series of actions: chasing up references and page numbers, proofreading (checking spelling), going to the library in order to look things up, looking at the catalogue to know whether to look things up. The processes of thinking and the products of thought had seemed to him at times to be at loggerheads. Even when that was not the case, it was still true that the sentence had been writ-ten to support the paragraph, the paragraph to support the chapter, the chapter the book and the book the *oeuvre* and the *oeuvre* to the contribution he had wished to make to changing the way his fellows, or some of them, might think or even behave. Such tiny steps and such a long journey! The pilgrimage across the world to give an hour-long lecture, to spread his cognitive seed on frequently infertile ground, provided an uncomfortably explicit representation of the permanent entanglement of his agency in the Kingdom of Merely Intermediate Ends. There was always another goal transcending the stated aim of an action. As he ascended the hierarchy of aims in his thoughts, and goals became more general, so banality threat-ened: 'my responsibilities fulfilled, a secure future, being loved and respected by others, enduring happiness'. Or, to borrow the moral philosopher's cliché, 'flourishing'.

Hence his occasional rebellions against the insatiable demands of (seemingly) serious purpose. Towards the end of his life, RT had felt more sympathy for those who had devoted themselves to pursuits that his younger self had dismissed with contempt. Those who built battleships out of match stalks, or etched the images of all the known saints on pinheads, who dreamed of completing a collection of stamps or train numbers, who edited over a decade the works of a justly forgotten Latin poet, who devoted every free moment to trying to achieve a golf handicap in single figures – perhaps they had a point. The pursuit of pointless perfection – just as much as the great, important tasks of life – bound days together, supported a narrative of progress (each day nearer to the full set or some other measure of completion) without the pretence or illusion of adamantine meaning that his corpse now exposed for the illusion it was. The hobbyists had constructed a world within the world, as a way of making the world their own. This microcosm was rendered more secure by a common cause, expressed in clubs, societies, and communities, with others similarly afflicted with the same almost-innocent passions.

'Almost', because they had turned their indifferent backs on the world beyond their narrow pleasures, and foresworn the Noble Aim of Leaving the World a Better Place. Outside of the calm, cosy bedroom where match-stick was glued to match-stick, the hushed library where footnotes were perfected, or the genial clubhouse where tee shots were re-examined one by one, hungry children howled, slaves died building vanity projects for millionaires, and the destitute perished of hunger. Thus had he argued with himself for and against the Pursuit of Pointless Perfection.

Behind the cultivation of seemingly futile activities was, perhaps, a sense, usually suppressed, that no final goal was ever reached; that all effortful business served only intermediate ends. Might it not have been true that, inside every trainspotter, was a suppressed tragic sense of life? Or, when it was a question of experience for its own sake, had he not felt that he had never fully experienced anything, because he could not (to take a random example) walk the streets of *the idea* of Paris. And so, like his fellows, he was always on the move, rolling out project after project ahead of himself, forever trotting his small portion of the globe.

And it may be here or hereabouts that we might find the place of art in the life that has now ended. If art was such a *serious* pleasure, inseparable from his sense of his life as a quest for a significance that went beyond the quotidian concerns that stifled or hid the message of the stars broadcasting just how little that life was, it may have been because it seemed from time to time to offer the possibility of rounding off the sense of his world, lifting him out of the kingdom of intermediate ends. Music, painting and literature spoke to consciousness conceived as an end in itself but, unlike other necessarily transient pleasures, seemed to dilate into those great spaces opened up, but also emptied, by knowledge.

If art had represented one pole of the ado that filled his life, the activities of daily living, necessary for physical and social survival, represented the other. Think of the doings – increasingly obtrusive as he aged and started dying – that attended his passage from the night's sleep to the day's wakefulness, a passage that had been repeated, with variations, over 25,000 times.

Preparing and eating food and drink (with all the subsidiary actions around setting the table, washing the dishes, getting to a café, and so on), personal hygiene (from brushing his teeth to taking his suit to the dry-cleaner), and emptying his bladder and bowels in the right place had occupied much of his time on earth. He had once calculated that he had cleaned his teeth over 50,000 times and that the noodle of toothpaste he had squeezed out of constipated or over-compliant tubes was around 700 yards long, sufficient to connect the bathroom with the newsagent's or to circumnavigate his house and garden several times, with eras of his life as a brusher of his own teeth marked by the transition from pure white, to the popular red-white-and-blue striped (a patriotic technological miracle that was intended to amaze and did) to the single colour, crystalline blue. These were not epochs a developmental psychologist or a historian would accept but a periodization of sorts.

The spectacle of his naked body in this his final bed, covered by, rather than clothed in, a sheet, brings one activity above all to mind: dressing. At the two ends of his life, he had been dressed by others but for most of the interval between nappies and continence pads he had dressed himself. To say that this reflexive activity – with the seemingly straightforward aims of protecting himself against cold (and sometimes heat), against wet, and (since it presented him as a person with a carefully calculated appearance rather than an organism) the glances of his fellow men – was complex might be something of an understatement. Finding, selecting, picking up, and inserting himself into his clothes, and fastening (buttoning, zipping), and tying, drew on a multitude of cognitive capacities (perception, memory, recognition) and motor skills (balancing on one leg, gauging the relationship between an arm and a sleeve).

Each item had demanded something slightly different of him. Take the sock. He had to remember where he had dropped it the previous night, peering back through hours of dream-stained obnubilation. Then there was the challenge of recognizing a soft woollen ball as the left sock fused with the right sock, with insides and out-sides invaginated into each other beyond the reach of any diagram. His motor imagination was then required so that he could foresee the consequences of unpacking the ball, straightening out the two protagonists, and inserting each naked foot in turn into a cul-de-sac shaped by the very item muscling into them. And this was merely preliminary to the shoeing of his stockinged feet. Assisted possibly by an index finger acting as a shoe horn, his feet were inserted into the twin (silent) screams of elliptical right and left orifices fashioned out of canvas, rubber, or leather, thus gagging their silences. Against such achievements, it was hardly surprising that the additional skill of grasping the spatial logic of increasing the thickness of sockage to fill out larger-size shoes (or climbing boots), not to speak of his mindfulness of the constraints regulating the kinds of socks to be worn with what kinds of shoes, were hardly appreciated, especially given their effortless deployment. Or the personal organization (tidy-ing up, storing, shopping) and social organizations (shops, transport systems to bring the goods to the shop, and the customer to take them from the shop) that made it possible to incorporate the foot-wear into an ensemble that also included underwear, shirt, trousers, ties and cuff links and shoelaces.

Competence in dressing had helped to define 'key stages' in his development from infant to independent adult – and, alas, also landmarked his disintegration. Because it involved virtuoso per-formances such as pulling a pullover over his head without losing his balance in the temporary darkness created by this act or tying his shoelaces (balance, dexterity) or fastening a tie in the fashion required of the moment (so much being signalled by loosening) or linking cuff links – particularly difficult on his right side when the non-dominant hand had to lead – it was an early casualty when agency started to fail. Beyond learning how to dress to conserve heat, provide comfort, and maintain propriety, he had acquired the rudiments of a 'dress sense' of what went with what, of what uniform was worn on what occasion, of the right level of *habillé* and *dishabillé* to ensure that one looked smart, cool, respectful, casual, formal, as required. The evolution from the calculated scruffiness of his youth to an automated (if at times slightly resentful) conformity to suit different occasions paralleled his thickening CV and the consequent diminishing importance of the current account of visual appearance relative to the deposit account of office, standing, and responsibilities. His dress sense was an aspect of an exquisite social awareness that an ethnomethodologist might have astonished or bored him by spelling out.

Notwithstanding the wizardry of motor skills and attunement to the social register of occasions that dressing required, RT had, for most of his life, performed this action on autopilot while thinking of other things. And so he could angrily struggle to insert a socked foot into a shoe (open-mouthed like a fledgling whose entire being is poised between a plea and a demand), while simultaneously worrying about a patient he was due to see later in the day, saying, yes, he was coming down for breakfast, and attending sufficiently to breaking news about the latest developments in the Middle East as to be able to discuss them with others later in the day through which he would make his socked and shod journeys.

These activities of daily living had been in part taught and in part self-taught: the basics were taught; the fluency was self-taught. And this was true also of many modes of ado. The range of his learned skills was breathtaking – from riding a bicycle, through knowing how to use personal pronouns or to complete sentences at the end of a thought, to tell the time, read, add up, use a map, show sympathy in a convincing way, when to stop talking, praise others without gush, repeat a joke, make sense of an electrocardiogram, take a patient's history, listen intelligently to certain kinds of music – though his level of attainment had been somewhat variable.

Fluency had freed him to make fun: life was not always serious. He was frequently cracking jokes, taking the piss out of his actions, himself, his conspecifics, his world and his others. He repeated the jokes that worked, with diminishing returns. The advantage of verbal humour was that it required no equipment – apart from a sense of what was and was not funny. Regarding this latter, he had no idea how niche his humour was. From an early age, his speech was littered with jokes against speech: his self-mocking mandarin verbal style was punstriped. His voice was overlaid with many others. This sense of fun, that was remote from the life-threatening pranks of boy racers and any impulse he might have had to subvert aspects of the accepted order, left everything largely intact. Horseplay was a minor presence in his life, sprees were rare, and jinks tended to be of a medium altitude. He did many jokey things without being able to say precisely why he did them – or why he did them in that way. A couple of times he enacted the preening gesture of 'shooting his cuffs', probably because he liked the thought that something that had once (perhaps in the Roaring Twenties) seemed desperately stylish was now quaint. It was another marker of the many and curious distances between the energetic RT and the still item under the sheet. As was his buying socks in order to amuse others: socks of outrageous colour, with lights that went on and off, or that played the 'Internationale'. To have been a member of a species that reported itself as laughing its socks off at laughable socks was to be a complex creature indeed.

As was true of all instances of *HS*, RT's competencies had been islands in an ocean of incompetence – hence the division of labour. His attempts to learn languages other than his mother tongue were patchily successful; drawing, playing, reading and writing music were beyond him, notwithstanding his passion for seven centuries of the classical repertoire. He would have made a hopeless engineer, a dangerous surgeon, a cowardly soldier, an indifferent orator. There were words he could never pronounce, no cravat was ever properly tied by him, and he never quite learned to gauge how long to boil an egg to transport it to the right state between slobber and rubber. Ballroom dancing – where courtship and military discipline, Mars and Venus, strictly and come were united – was a potential source of humiliation he learned early to avoid. His clumsiness was multi-dimensional though never oafish or loutish.

Many of the things he could manage were executed poorly. He had not infrequently fumbled and bungled. There was also the rummaging which was often unsuccessful and made him irritated. In part this was because of his skimpy approach to preparation and his capacity for boredom, that made him slapdash in getting together the wherewithal for whatever had to be done. He had done more than his share of ferreting and fossicking (though rather less than Mrs RT whose large, overladen handbag had meant that fossicking – for example, in urgent pursuit of a ringing phone, hoping to catch it before it rang off – was often frustrated in its aim) and resented both. He had spent far too much time trying with variable success to unjam the jammed, untangle the tangled, and translate printed directions into action. Admittedly, as he grew older, he had been less inclined to get cross with objects (and, indeed, with people), though he sometimes felt that he had more reason to do so, given that being a human agent was such a fiddly business, with picking up and unpicking, threading through, knotting and unknotting, and squeezing past after reaching down, pinpointing this and that, and so on occupied far too much of his infinitely precious time.

If his life had been at least averagely successful, it was because he had found a niche for a mixture of capacities and incapacities that created a customized definition of success which enabled him to be more at ease than he should perhaps have been. The idleness that meant he was far too content to keep certain incompetencies intact could be managed, though this did not fully explain why cushions in his vicinity were safe from plumping even when visitors were expected.

Most of his doings were 'joint and several' (as the law would describe responsibility for them) and the 'several' sometimes amounted to a cast of many thousands, most of them unknown to him. Consider the bread his salary had enabled him to put on the table and butter at one of his life's estimated 15,000 breakfasts.

Let us for simplicity set aside the kitchen (builders, fitters, and manufacturers), the table (painters, carpenters, sawmill workers, woodcutters, foresters, seed merchants), the crockery (ceramicist, potter, clay salesman) and the knife (steel worker, cutler), the lighting and heating, and focus on the slice of bread. What entered his mouth had arrived there via a network of long and circuitous journeys, encompassing agriculture, transport, processing, and marketing; and shaped by countless customs, rules, laws, standards, institutions, and modes of expertise. Prior to the interval between seedtime and harvest there had been the work of the farmer acquiring, inheriting, expropriating his patch of land, preparing it for sowing, purchasing the seed corn. Afterwards, there was the journey to market – whatever form this took – and the further journey to the flour mill. The next stop was at the bread-maker's – an individual or an industry – who oversaw the convergence of other elements (salt, clean water, wrappers, etc.). Each of these, of course, was another point of convergence. The ornate wrapping around the sliced bread required pulped and sliced trees, grease from a whale, various dyes and the skill of the graphic artist. The relay race continues with the transport of the loaves to the shops where the bread is sold. From this stage in the journey we may unpack many things – the purchase and distribution of petrol, the training, hiring and firing of lorry drivers, all with their dockets, bills, and schedules, the manufacture and maintenance of vehicles, to mention a few. And RT himself can add his own journey to the shops where the bread is an item ticked off a list. The return leg delivered them to the kitchen, the stage on which bread, knife, butter, table and plate had met. And finally, his hands washed, he raised the slice to his mouth and bit into it.

Similar complexities attended other direct means to survival – drink, clothes, and shelter. We have already reflected on the design of reservoirs, of water pipes, of taps, and of glasses, and the systems to ensure that the water was uncontaminated. We note here only that it required as much material, cognitive, and regulatory infrastructure as ensuring the energy intake necessary to keep us running, breathing, talking, and to maintain the fabric of that which runs, breathes, and talks. The complexity of clothes – with their multiple functions (warmth, protection, decency, status, signals – which were sometimes spelt out: T-shirts like billboards, dresses that were mere podia for their labels), the parsing of elements to make an overall grammatically pleasing appearance, and the relationship to time, culture, and climate – make the frocked and besuited into individual nodes in an unimaginably complex network of networks. As for dwelling places – and the boundaries intermediate between clothes on the inside and national borders on the outside – they are many-layered and each layer is many-layered.

Much of his ado was paid work. More ado was duty outside work. There was general togethering which didn't fall entirely inside or outside the category of work or duty. And some ado was recreation – play, fun, mucking about, pottering.

His preferred modes of recreation had been rather quiet, solitary, dual, or in small groups – reading, listening to music, walking, immersing himself in cities, villages, countryside. He travelled to many places simply in order to transport his body to places to enjoy experience for itself; for the sake of what he would see, hear, smell, taste, and touch; for a different language and culture; to translate expectations built up by words, images, and rumours, into a reality that was never, of course, precisely congruent with them. And there was the joy of journeying for its own sake: miles under the belt, a sense of achievement, the complex, mixed pleasure of testing himself against mountains, and waves, and footpaths. Behind it all was the fundamental sense of getting further away; of the enchantment of distance, of which we have already spoken.

Thus the pastimes that helped to pass the time leading up to this stillness under the sheets. He had joined the human race at a period in history when it had become obsessed with sport, an obsession force-fed by mass media whose presence increasingly approximated to ubiquity, something he had resented. His early experiences of compulsory sport – and his limited competence at running, jumping, kicking, handling, batting – made him wary of taking games too seriously, except when his children were involved, though it was their hopes, expectations, and disappointments rather than the events they participated in that mattered. Professional sports may have been superior to playing about (the World Boxing Championship to pillow fights), to horseplay (Badminton three-day eventing to chasing each other with snowballs) or even to mucking about, but they were still fatally useless. Which thought provides an *entrée* to a rehearsal of one of his most practised rants, by which we might honour his memory.

All games, he had pointed out, were zero sum or worse. The fulfilled hopes and aspirations of the few, he had argued, are built on the disappointment of the many – leaving aside false consolations such as achieving a 'personal best'. The inverted pyramid of disappointment, with each layer of success being raised on a fatter layer of failures and dashed hopes, underlined how the passage from scratch cricket to the Ashes, or from games in the playground to the Olympic Games, was an exorbitantly expensive journey, costing vast quantities of blood, sweat, and treasure, that was not worth the candle. Sport left no lasting legacy which is why it was always looking towards the next series, the next season, the next Olympic Games, thus emptying the present with an increasingly assertive future. This was but one of the many reasons why he resented the unremitting torrent of sporting 'achievements' that demanded more and more of his already too divided and distracted attention. Why, when the nation rejoiced over, felt proud of, was cheered by, some triumph of a plucky British victor, he felt no joy, swelled with zero pride, and was resolutely uncheered. He thought instead of the pointlessness of it all.

The fat past of opportunity cost, the thin contrail of present success, and the obsession with the future, was an involuntary metaphor of the futility of life. Every victory was merely a signpost to possible future victories; all successes were waiting to be made into failure by the future successes of others. On the imaginary line that linked participation in the game of peep-bo with playing in the World Cup (with its steep upward gradient of preparation, kit, resources, training), he was content to remain close to the beginning, where low-cost events such as throwing a stick for a dog or playing bat and ball on the beach were located.

Even so, games had occupied a good deal of RT's waking consciousness that had ended in this similitude of sleep. Hours jumping over a bar in a wasteland, a few years when he played cricket, football, had wrestling matches, cycling circuit after circuit, accounted for much of his early life outside the school room or domestic duties. And games were an ideal way of togethering with his children. And the games that required his time went beyond the track, field, and the pitch, to the board. He underperformed at Scrabble, chess, quizzes, if only because he was too aware that there was (by definition) something not entirely serious about games.

On those rare occasions when he participated with more enthusiasm than he felt he should, it was because what happened could always be translated into the purest form of wealth: numbers. There was the achievement of jumping higher, running faster, piling up the runs, accumulating the cash on the Monopoly board, and so on. The formalization of play into sport by the formalization of rules, the formation of teams, leagues, and championships, the professionalization of everything that looked like a game, the primary discourse about matches, and the secondary, tertiary and quaternary discourses about players, managers, referees, owners, sponsors, and what they did and didn't say, and rules about the application of rules, could be redeemed a little by the statistics that gave narrative-structure and dignity to games, generating the virtual outcomes that terminated in the cenotaph of almanacs.

This, and the fact that informal playing could soon deteriorate into bullying, quarrels, and teasing, went some way towards justifying the calcification of *Homo ludens* into the professional sportsman, coach, sports administrator, and even the paraphernalia of seriousness that made it perfectly OK to spend more money on sport than on feeding the hungry, or for hundreds to die building a stadium in a country that had bribed its way into hosting a prestigious football tournament that would, it hoped, distract from its iniquity.

There were long intervals when ado consisted simply (and nothing could be less simple) of being together with others, engaging in activities in an indeterminate zone between playing and working. Gassing, chattering, greeting, listening, pretending to listen, had occupied countless hours. In addition, there were many things that might not seem worth reporting or recalling; for example, picking up bits of paper off the floor, dealing with spillages and interpreting and attempting to obliterate stains, flicking absently through magazines, reading instructions with care or carelessness and irritation, angling a sheet of paper to catch the light so that he could watch the ink dry before folding it in two, looking for a particularly smooth flat pebble suitable for an attempt to break his personal best for ducks and drakes, reporting that something was not working properly or noting suspicious activity, or complaining about being endlessly requested to do so by public announcements in the latter half of his life. Outstaring and earmarking likewise.

There were doings that, though not highly esteemed, were remarkable, if only in virtue of being christened by turning extraordinary mass nouns (or massless nouns) into verbs – for example 'dusting' and 'vacuuming' – making them stranger even than 'mopping' (wet floors), 'mopping up' (the remainder of the enemy forces) and 'sopping up' (gravy with bread, spilt drinks with napkins). And there were items that attracted verbs without fully qualifying as actions: bracing himself for a loud crash; so ordering his affairs that the edge would be taken off his or someone else's ire; outstaring, earmarking, stipulating, and claiming to have stipulated; suppressing a smile; seeking a stranger's approval; disliking his mirror image. There were many modes of inspection, though they were not of equal standing: inspecting the back of his hand did not earn the kind of mention owing to inspecting a racetrack prior to an important race, a school in special measures, army kit (properly cleaned, polished and ironed), or a passenger's ticket, or those things that fill the working hours of a *detective* or a *chief* inspector. Allowing himself to be distracted by the sound of dripping, refraining from wassailing, and admitting to being disinclined to do this or that, seemed even less qualified to be included in the ado that he did. Mooching, hanging about, skulking, finding himself at a loose end, feeling self-conscious, getting confused over the Jutes and Hengist and Horsa, deciding not to engage in a dispute over Hindu doctrine, vocalizing exasperation when a tick jumps out of a box (thereby proving to be a jack in a box), and a task has to be repeated, or when folding a tablecloth (a complex feeling distributed between the cloth, the person who asked you to do it, the custom that requires it and the seemingly deliberate dumbness of fabric subject to the laws of mechanics), were definitely in the no-man's-land between activity and passivity. He would, if challenged, sometimes locate them along with ill-defined actions like pampering someone, schmoozing, minding the gap in response to being reminded to do so, being willing to listen to 'a big ask' or even catching sight of his own oddness by deploying his gaze; and at others times he would locate them on the side of passivity along with being intrigued or (to use the customary exaggeration) being mesmerized or 'absolutely' mesmerized, being called to order by a rung bell, a knife on a drinking glass, clapped hands, or being sandwiched in a scrum, on transport, or in a crowd.

The question 'What do you do?' most often meant: 'What is your occupation?'. For many years, his waking hours had been devoted to sustained, paid, and contractual, highly structured ado that drew on a multiplicity of skills, know-how and know-that. His time, like that of candle makers, bricklayers, lawyers, footmen and handymen, punkah wallahs, professional vexillologists, locksmiths, weavers, toy designers, water sommeliers, chefs, goatherds, pest controllers, surveillance camera ombudsmen, data-entry clerks, leather dyers, infantrymen, dubbing mixers, sub-mariners, graphic artists, computer programmers, and teachers of history and of cake decoration was in the keeping of others for a certain number of hours per day or week or year. He was at beck and on call and his movements added up to 'offering a service', 'practising a skill', 'professing a profession', 'discharging his duties'.

Thus chronically occupied and preoccupied with an occupation – in his case for 120,000 hours over his forty-year career of medicine – the balance between 'doing as he was told' to 'telling others what to do' shifted in the direction of the latter as he gained one of the most potent socially endorsed abstractions: 'seniority', a key strand in the braided narrative of his life. The burden of responsibility, the pressure of competing demands had left their mark on this body, though rather locally: the creases in the face were not entirely due to photo-ageing, or the hollowness of the eyes to illness.

Since RT owes his presence on these pages to his being an instance of *HS*, it would be a mistake to examine his years as a doctor in any detail. *HS* is for the most part not medically qualified. Though he sometimes reflected on all the movements involved in being a doctor: winding back a tape, inserting a stethoscope into his ear, helping a patient on to a couch, palpating a liver edge, staggering down a corridor transporting case notes, ringing up X, Y and Z, hurrying to prepare a presentation, crossing out an excluded diagnosis, interviewing a research fellow, writing, writing, writing...

Much of his doing was undoing – his own and others' doings: unpicking, unsaying, dismantling, retracing. These counter-actions were still doings. 'Unning' was as effortful and active as positive modes of '–ing', as became evident when the action in question was *un*wrapping – not always a happy experience, particularly as the power and precision of those now powerless fingers began to wane. The struggle with sachets was a case in point and offers an irresistable opportunity to rehearse another of his rants.

The preordained weakness in the plastic to permit its tearing increasingly eluded him, however he tried to compensate for the fading of his vision with ever harder peering of those now unseeing eyes. The tag that marked the border between the upper and lower portions of a cellophane jumpsuit enclosing a packet of sweets or a biscuit, the fissure between the layers of shrink-wrapped cheese, irritated him for all that they were technical marvels. He had had not a few ill-tempered tussles with sachets containing the ketchup, mayonnaise, or vinegar that he requisitioned to create stronger gustatory harmonies to waken a palate that had become hard of tasting. As he increasingly invoked assistance – scissors, clippers, cutters – the barriers, as if in response, became more ingenious. It was part of a wider problem of The People – usually older people disabled by sleepy fingers and preconceived ideas – Against the Wrappings in an arms race, fuelled by the needs of mass production and the threat of ever more ingenious and absurd litigation – which he and the People increasingly lost.

While he was often decisive, or at least unthinkingly fluent, there were many times when he was assailed by inertia or fatigue or found himself powerless. More interestingly, his actions were often hedged about with doubts, when he held back, standing idly by because the alternative seemed to be to jump ineffectually or destructively in. Even the most confident, single-minded, and purposeful lives are streaked with necessary hesitations. He who does not hesitate is lost for failing to look before leaping: in all but the most reckless, crass, bull-headed psychopath, second and higher-order thoughts sometime intervene between the first thought and the action it commends. RT's life may have been more hesitant than most; indeed, his hesitations were sometimes so pronounced they seemed to stand to actions as muscle cramp stands to voluntary movement. His general uncertainty was a wide territory bordering many other domains. Out of his hesitations – ranging from roadside dithering, or doubts over when to greet an oncoming stranger, or to tell that joke, when to intervene in an outbreak of public disorder, selecting medicine or biochemistry for a career, the endless agony of clinical decision-making, whether or not to support such-and-such a political party, or person, whom to marry or whether to marry the whom in question, or simply considering options – we may select one example (after some higher-order hesitation) on account of its sending out a low light widely diffused: whether to use 'plash' or 'splash' to describe the voice of an object falling into water. The choice of using or withholding an extra 's' touched on some quite disparate memories: reading out loud Evelyn Waugh's *Scoop* in the seventies of the previous century, 'falling about' laughing (laughter sometimes removing extensor tone from the lower limbs), rereading it and not finding it funny forty years later, the room (in a friend's house) where the rereading successfully got him off to sleep, and the talk he was in Cambridge to give. Such was the internal stitching of his life.

Some hesitations proved to be eternal. The things he did not do on a particular day, he did not do for all time. Of these, few were a cause for regret. The invitation to go wassailing was not repeated but it would always have had the same answer. He did not regret never snorting with disbelief, paragliding, or using 'Best of British' without scare quotes. But there were other invitations which, once declined, seemed like lost opportunities, paths not taken, part of the process that led to a buried life, or a life not lived. While he who hesitates is lost, ditto (as already noted) he who doesn't.

RT is no longer a striver of any stripe. Past is the brute effort of pushing and pulling, the precision endeavours of tying and untying, wrapping and unwrapping, not to speak of fumbling, fossicking and winkling, of squeezing himself into a car, a welcoming partner, an exclusive club. The effort, less visible but no less real, of learning how, and learning that, and practising and memorizing, closing the book and reciting lists to himself. And the effort, by dint of concentration and application, to perform well, to keep going in long hours of medical practice or child care, to remain awake, alert, to make sense of what is happening, to understand, to read a face, a tone of voice, or a difficult letter. Gone too, the resolve to overcome the tiredness of long effort, to meet the demands made upon others upon whom he made demands, to keep others happy, to control his irritation or impatience, to knot that recalcitrant shoelace, to insert himself into a pullover that presents itself inside out, to fiddle with the fiddly. Behind him the effort of doing that which it is necessary to do in order to make it possible to do what has to be done. Of progressing by endless indirection, through dense thickets of preliminaries, to lay the ground for actions that would pave the way to ends that were always only temporary.

All RT's passion – and action – is spent. The total of his agency, his minute tweaking of the course of events, is summed. Undone, he rests from the labour of being himself, in something that is less than rest, less than peace.

All ado done.

ten

A Space of His Own: Having

For seventy-five years he had hithered and thithered with varying clarity of purpose in a theatre in part constructed out of the natural world and in part by the community of minds to which he belonged. The stage was cluttered with props that had assisted his actions and transformed the world into a kind of home. He had been a man of moderate, that is to say, considerable private and shared property. And now it has come to this: between a mattress and a sheet, a naked unowning body. His body chaperoned only by a shroud that does not belong to him, is the starkest expression of a wider dispossession. That mother of all commonplaces, 'You can't take it with you,' is a necessary consequence of there being no longer any I to be a you.

From the standpoint of this absolute privation we can get some sense of the mountain of 'stuff' that had sustained him in his everyday life: the gear and tackle and trim, the kit and caboodle; the hundreds and thousands of items that had eased and amplified, simplified and complicated, his interactions with the world, and with his fellow creatures; the possessions that had magnified his presence; the necessities that had made his life possible, and the luxuries that made it more endurable or enjoyable; the things that had helped him to survive from one day to the next, insulating him from those forces that had fathered him but were careless of their children.

Though comfortable beyond the dreams of the wretched (the overwhelming majority of his forerunners and contemporaries), he had not thought of himself as wealthy. Even so, there is simply no *raison* powerful enough to bring all the items he had owned or used in his life, or even those that had still been in his possession at the time of his death, into a single *catalogue raisonné*. An exhaustive inventory of the clobber that had passed through or accumulated in an ordinary life such as RT's would seem like a *Whole Earth Catalog*.

It is natural to begin with necessities. For the greater part of the history of humanity, this would have meant very few items – a hand axe, a pelt, a cave of one's own, and perhaps private means for creating fire – and not even a handkerchief to place at the end of a stick to carry them. One epoch's luxury is its successor's necessity; and so the category of 'the necessary' had widened in the centuries before he was born and, at an accelerating pace, during his life. Very little of his estate had even a tangential relationship to organic continuation. Augmentation of the inventory beyond the basic wherewithal was an inexorable trend. And even the wherewithal had become absurdly complex.

Take nutrition. Nothing so simple as for cows who neither hunt nor gather, who put their mouths directly into their million identical courses of menu *formule* without the aid of cutlery or even their limbs. Consider just the later stages, after the food had been acquired – usually bought rather than caught. Knives, forks, spoons, plates, cups, pots, pans, exotics such as egg-slicers and whisks, ovens, microwaves, tables and table clothes, recipes and recipe books, were members of the cast of thousands assembled to support the network of activities by which nutriment was processed, distributed, and presented to RT's mouth – an orifice that has now admitted its last supper, breakfast, and dinner. The aftermath of eating requisitioned its own support system – the kitchen sink, the washing-up bowl, cleaning liquid, dishcloths and mops, tea towels, aprons – all designed to help reverse the disorder, a potential breeding ground for pestilence and the odour of rot, created by preparing and eating a meal. And both sides of the repast required hot and cold water, general and focal heating and lighting, delivered to the point of need, and fans and deodorants to remove the smell of cooking unwelcome to those who have already eaten or who are merely over-smelling someone else's meal. (The amount of disorder created by preparing and eating food was the obverse, perhaps, of the centrality of its role in providing the means to reverse the tendency towards dissolution in the human body. Dirty plates are entropy's revenge.)

The list is, of course, incomplete; and equally long and varied registers could be compiled for the support of other daily living tasks. Grooming, for example, upon which we have already touched. Water adjusted to the right temperature and available through a variety of founts and ports, soap, hand and bathroom towels, flannels, hair lotion, deodorants, combs (of diminishing interest as time had weeded his vertex), were all recruited to separate his body from its own products and from the small change of a world whose soil was re-categorized into stains and markers of potential infection, of poverty, of self-neglect and undesirability. His daily makeover was at least less complicated than that of Mrs RT. No lipstick, protruding like a bullet from a cartridge, eye-liner to put the gaze in italics, or false eyelashes played any part in his life. Even unisex cotton buds had only brief appearances, as was entirely proper since they could damage their favourite cranny, the ear canal, and he had rarely used them to apply cleansing alcohol to inaccessible parts of delicate machinery, such as the computer on which this sentence is being written. And nail brushes, files, scissors, clippers were available to both parties, though varnish and varnish remover were used exclusively by the distaff side of the family.

Urination and defecation – and the imperative not to wet or soil himself – had had their teams of mediators, assistants, and attendants. He was, after all, born at a time in history when there could be a debate about the respective merits of hard and soft toilet paper, when it was marketed on the basis of virtues – colour, texture, and information – that would have cut little ice before the Anthropocene. The splash and plash (*vide supra*) of successful urination and defaecation, and the urgent sound of the flush that tidied the result away, were some of the most familiar sounds in the *sinfonia domestica*. These were merely preliminaries to the day's business – work and play, getting and spending, earning his crust, contributing to manufacturing output, swelling the GDP, serving others, making his way in the world, cutting a figure, or even what he liked to think of as 'making a difference'.

The line dividing necessity from luxury was blurred not only because of the already noted trend whereby yesterday's luxury becomes today's necessity but also because even those items that could be classified as necessities were elaborated in a way that could only be described as luxurious. He had relished this wonderful deviation from functionality.

One of his favourite examples was the ornamentation of the wrought-iron tables in certain pubs, in which there was a face carved at the top of the legs and the same legs had feet shaped into the paws of beasts or were even more exotically ornamented. He had not infrequently exchanged glances with the king of beasts, when, awaiting a response to his knock at a neighbour's door, he examined the beautiful evocative, leonine sun adorning the knocker in his hand. The lowly tin mug did most of what was required of it. Nevertheless, the salami division of drinking water into drinks – allowing a continuous flow from stream or tap to be widely, justly, or conveniently distributed – licensed numerous developments that branched off in a multitude of directions. Delicate, porcelain teacups decorated with scenes of appropriate *gemütlicheit,* or mugs as merchandise proclaiming allegiance to causes, or boasting their place of origin (exotic compared with their kitchen destination), or the tourist traps they commemorated for those who had travelled to them, or as tepidly appreciated gifts for those who had not, were just a few examples out of the catalogue. Curtains required swags to maintain, during their periods of rest, a stately procession of marmoreal folds rather than a noise of creases, and this provided another opportunity for ornamentation as a particularly fine pair of brass ones had reminded him in a hotel in Lund, Sweden. His brogues had an exquisite pattern for the sole purpose of proclaiming themselves as brogues and there was no shortage of ways of exploiting a pleasing contrast between the colour of shoes and their laces, while buttons, bows, and even bells, were just a few of the possible adornments. Hats of uncertain purpose were adorned with tassels or a pompom that marked where the wearer came to a full stop or a dot on the 'I'. The belts that held up first the short and then the long trousers of his childhood, boyhood and early youth, were buckled by interlocking snakes. Ordinary edibles were presented in wrappers that carried evocative views of their places of origin. Cows, smug in the knowledge of a life well spent in the service of humanity, were frequently portrayed on the milk products whose fat content may have hastened his arrival at this place of total dispossession. All goods, in short, had their deluxe allotropes and every surface was an opportunity for aesthetic expression. While the line was drawn at gargoyles on coat hangers, the principle was that, when it came to artefacts, humanity rarely just walked when it was possible to dance.

This was most evident when it came to the means of sheltering or protecting him from the many threats – to life, limb, and comfort – that the world presented. The inmost shelters were the multiple layers of cloth in which he swaddled himself. RT was perhaps less preoccupied than many with the contribution of haberdashery to modulating his presence in the world; even so, adjusting his clothed appearance to the circumstances in which he was to appear was time-consuming. The rival claims of boxer shorts and Y-fronts to the office of concealing, protecting, warming, and otherwise comforting his more intimate parts, did not requisition much of his attention. But the more visible layers – the shirt, the tie and the pullover (like bags called 'grips' named according to the means by which they could be brought into action), trousers, jackets, coats, scarves, hats, and shoes – did concern him above the minimal necessary to achieve the appropriate level of formality or informality, in order to be uniform with others where it was expected, to be unconventional according to the demanding conventions of unconventionality where it seemed right, and to be padded and wrapped up or lightly clad, as the weather and the activity to be engaged in required, to be kitted out where kit was needed. The Wednesday evenings when as a teenager he laid out and prepared his uniform for the following day's military exercises underlined the convergence between equipment and clothes and how the ironed, blanco-ed, Brasso-ed, polished outfit was simply an ensemble of symbols. Boots – the metonyms of military bio-ware ('boots on the ground') – existed to shout in chorus as the ranks snapped to attention and, by shining, to testify to the discipline of the wearer and the military capability of the troop of which he was a part.

His dress sense, though less developed than most, proved on close inspection still to be exquisite, bearing striking witness to advanced cognitive capacities and social awareness. He may have tolerated odd socks, tramlines on trouser creases, ties that quarrelled with shirts (and were sometimes a meeting place for the art of the chef and the skill of the couturier), but there were still modes of dress that he resisted. Thin, sleeveless pullovers (to single out one example) he believed to be charisma-quenching and overly domesticated. He had a definite, if idiosyncratic, grasp of the concept of 'the natty', especially as applied to jackets. In general, however, any fashion statements he made were inadvertent or *sotto voce* and contributed to his generally low visibility among strangers. Like most of his fellows, he had cut more of a dot than a dash. Even his blunders were low-key; nothing as crass as combining Bermuda shorts and thermal underwear.

His mode of dress had to be adjusted to suit the phases of his life, the fashions of the decades, the cycling of daylight and the circadian rhythms of consciousness (sleepwear and wakewear), the relative position of the earth and the sun (his seasonal collections), and the many different situations in which he found himself. The sum total of haberdashery that had passed through his life, and temporarily lodged in wardrobes, cupboards, and drawers, and on shelves, in cloakrooms, on the hooks here and there, in suitcases and bags, and stored in attics, would have astonished, perhaps shocked, him if it had ever been laid out in full. The row of jackets, the multitude of hats worn in a largely hatless epoch (fedoras as well as caps whose brims and peaks stood in for hands shading his eyes from sunlight, the cranial coating of the woolly hat and the head-sock of the balaclava), the ruined pyramid of socks, the snake's nest of the tangled ties, a multitude of scarves, was an involuntary history, if not of something as definite as tastes, but at least of periods of his life when they had been bought, worn, and even, for a while *sported* with mild pride. Periodization was somewhat muddled because items that went out of fashion might have been retained for private – or a different – use. Retired trousers were re-hired in less prestigious roles: digging rather than dancing.

Because he was not the tramp of the comics he read as a child who wears his entire wardrobe as a means of storing it, clothes on active service were outnumbered by those not currently in use and they in turn were outnumbered by the disused. The latter, as well as seeming to be aspects of his past self that had gone into cold storage, were also reminders that he could not animate all his possessions with attention, even less activate them with use, an observation that carried the larger message: at any given time he was smaller than his life. He had found this somewhat disturbing but not as disturbing as the kind of thoughts that might assail him in the bedroom on a sad, but ordinary, Wednesday afternoon. Then the emptiness – limbless sleeves, torso-less breasts, and headless collars – of his uninhabited clothes warned him of a future when his body would no longer inflate and animate any of them; of his present state when he had undressed for the last time and *all* his clothes were empty of him, as he lay naked under a sheet.

While his clothes were the most intimate of the shelters that had protected and comforted him during his stay in the world, the most prominent was his house. His place of abode, relatively fixed, though not by standards set by mountains and other fixtures in the cosmos, was not only the thickest layer of the exoskeleton that enclosed him, it was a pocket of space to which he had privileged access, from which the world at large was excluded, except by his invitation, and to which letters (personal, financial, professional) and phone calls had been directed, and were arriving even now, addressed to a name that had lost its referent.

This had been the capital of Here, the most important of the many centres in his polycentric world, the most trammelled of places, where the ley lines of his presence had converged most densely, for it had been in this building, in these rooms, where he had been most a regular. It was around this 'heremost' that the universe was arranged most authoritatively into 'nears' and 'fars', whence 'away' gained its degrees of remoteness, where his days most often began, and where they most often ended, where his journeys started and where his journeying ceased, as the hunter from the hill, and the sailor from the sea.

And it was here also that his possessions had been assembled in their largest and most varied crowds, and piled most high. Yes, there were other gatherings – in his office, for example, and, for a while, in the attic of his parents' family home, and the odd locker here and there – but this was the epicentre of his ownership, the multiloculated capsule that most frequently enclosed his hours, his activities. It was the theatre of more of his struggles and joys, the stage of more of his conversations, than any other.

Home ownership was threefold. He possessed: a certain amount of space; the material (bricks, mortar) that enclosed that space; and the items in the closed space. More strictly, his was co-ownership. There was shared entitlement to the address and to an informal muddle of possessions with his spouse and children, a casual order that was the opposite of the contractually defined arrangements that prevailed in the outside world. At his front door, contract gave way to covenant and costs were less pusillanimously counted.

The over-arching possession that was his address had numerous components – most obviously the fixtures and fittings that adorned, modified, and beautified the fundamental structures. Like its owner, the house was clothed: carpeted, curtained, and papered. They softened surfaces whose primary duties were to be firm, hard, resistant, reliable, and permanent – a stable platform for daily life.

Of the large and small possessions that were contained in the receptacle of his home, the biggest beasts were 'consumer durables', some of which stayed with him from early in his adult life, when he had first had an address of his own, to his very end. They outlasted him, and were available even now to be inherited, sold, dumped, or chopped for firewood. The sofas and the armchairs, tables, beds, sideboards, wardrobes, toilet bowls, sinks, were such familiars as to warrant little additional notice in his daily life, other than that which was dissolved into his absent-minded interactions with them. The blue-printed trade mark on the washbasin, overflown by jets of toothpaste-scented saliva, the screws securing joints in cupboards, pelmets elaborating the curtaining of windows, and hooks on the back of doors received even less attention. To be at home was to be only palely aware of what surrounded him. Here was where things could, for the most part, be taken for granted.

Lamps and (occasionally) candles, and gas fires breathing flames, bars of electric fires holding their glow to themselves, and the flameless warmth of radiators, had been able – with the assistance of windows and doors that could be opened and closed and bijou breezes-to-order issuing from fans – to maintain a constant lit temperate zone irrespective of the weather outside or the relative positions of the sun and his patch of the earth. A flick of a switch countered the darkness and cold of outer space. In a universe of violent extremes (though it knows nothing of its violence) the domestic sphere had been an underappreciated haven with the temperate being dismissed as tepid and the regulated as dull.

Fixtures and fittings were vastly outnumbered by the other items in the thousand-nooked treasure chest of his fixed abode. Tracing even a single motif round his house would have exhausted his descriptive patience. Consider (again) glass, with its magical property of being as translucent as water but as firm as baked clay, as clear as ice but obdurately unmelting. It was everywhere: in the windows that allowed his gaze to look out while keeping at eye's length the cold and rain of the outside into which he was looking; in lamps, where it was fashioned into the cornea of a one-eyed torch, the transparent peel of a 100 watt bulb, and the frozen spray, arrested in its fall, of a chandelier; in vessels which permitted mass-noun fluids such as water, beer, and wine to be divided into count-noun drinks fashioned into geometrical shapes. Glasses ranged from the merely utile (the bearers of aliquots) to the vulnerably beautiful (adjectivalizing towns and cities where craftsmanship, artistry and mystery met – 'Venetian' etc.); from the sherry-bearing thimble to the ale-bearing yard. Lenticular glass enlarged the very small (revealing worlds within worlds), reeled in the very distant (weaving light from remote places into images of itself), and, supported on his nose and ears, sharpened the borders of a world populated and defined by edges. Sheeted, it protected the mirror's silver designed to create a virtual en suite and replicate the rooms and the *verismo* micro-dramas within them in the form of weightless images. Glass also protected the clock face whose hands orchestrated the activities in the kitchen, clenched into a paperweight etched with the image of a foreign city, and permitted observation of the carousel of food cooking in the microwave. Deftly shattered, it was the stuff of marbles, beads, necklaces, and jewels. And it lay at the heart of the camera that gave the visual skins of moments a taxidermic afterlife in which the snapped and shot could be looked at indefinitely. There were variants of glass – frosted, polythene – that separated the transmission of light from the possibility of detailed visibility, an effect echoed in the natural polythene of mist depositing itself as condensation or ice on windows.

The spirit of glass was present in many other kinds of stuff; for example in that marriage of windows and paper called 'cellophane', diaphanous cellulose enabling the wrapped to be multiple and yet unified, protected yet visible. The windows in envelopes allowed the name and address of the addressee to be communicated to the postman, while the contents of the letter within were kept confidential. Such were the bespoke filters that regulated the penetration of the public into RT's private world.

His habitat was a hoard of artefacts. From the toothbrush to the secateurs, the fish-slicer to the radiator key, from the calendar to the lawn mower, the car jack to the pillow case – they all had their moments of indispensability. And how numerous they proved to be! Towels that lifted films of water from faces, feet, crockery, cutlery, floors came in fifty-seven varieties, such as 'tea', 'bath', 'beach', 'hand', and 'flannel'. Mats – 'door', 'bath', 'place', 'beer' – likewise. He was copiously kitted out with items to assist him when, as happened not infrequently, it fell to him to divorce connected items or divvy up continuous stuff. Cake knives, bacon slicers, carvers, mincers, dicers, graters, peelers, scrapers (that manufactured the informal peel they peeled), scissors, sharp edges cunningly built into cling film dispensers, were just some of the large cast of tools tailored to assist his dissections, apportionments, and distributions. The divisions were sometimes quite esoteric; for example those made possible by lemon zesters, separating liquid juice from the multiphasic whole of the citrus. Others broke down divisions more radically – as in the case of mashers, pulpers, whisks, and blenders.

Divisions in other domestic spaces, in the garden, in the world at large, all requisitioned their cognate instruments to permit cutting, sawing, digging…The list is not endless but too long. The same could be said for those items that helped him to groom surfaces disfigured by stains, spillages and dust – mops, dusters, vacuum cleaners, flannels, oil rags, squeegees, baby-wipes (this last was frequently requisitioned for purposes that exceeded their wilfully mis-read job descriptions, being mobilized long after the last baby had gained sphincter control and no longer befouled the nest it had fled for tertiary education). He had an affection, nevertheless, for certain stains, and blots, and scuff marks as mnestic markers of past use and of the past when they were used, a subtle moss testifying to days outworn – imperfections underlining his ownership exercised through using, handling, or merely rubbing along with them. It even gave him pleasure to imagine the delicacy of the processes by which earth filings, so finely divided as to be largely invisible, were deposited with such equity from the air freighted with them on the surfaces dusted in the hebdomadal cycle of house-grooming.

The toolbox was supplemented, and even superseded, by those semi-autonomous tools called machines. They spared him effort – the grunt of pushing and shoving, the grind of repetition – and enhanced his powers. The number of machines upon which he depended multiplied as the years went by – to the point where, as we have noted, he visited other machines – treadmills and exercise bicycles – to engage in elective arias of effort to compensate for the loss of the immemorial recitative of involuntary grind. Because they were energized by gas and electricity and were able to ape fragments of intelligence due to the thousands of patented ingenuities within them, a press of a switch was sufficient to mobilize their assistance. It was possible by the twiddle of a knob to calibrate the quantity, or determine the type, of help they offered him, or the intensity and duration of power exerted. This use of dials to translate positions into instructions – a manifestation of the extraordinary cognitive capacity of his conspecifics to think across categories that had little relationship to sectors of space – was overlooked because so commonplace. Except, of course, when he was faced by machine failure and he had stared at the device in blank incomprehension seasoned by rage.

Machines changed his and others' lives. In due course they became indispensable: more instances of 'bare necessities' being endlessly defined upwards. Living without a mobile phone or access to emails – technologies unheard of before his middle age – was in his later years regarded as wilful or eccentric. There was a moment when what he had hitherto called 'the phone' was re-named 'the landline', when meetings were specified as 'face-to-face' as if that were a rather special sub-group of human encounters. Speech had not yet been rebadged as 'throat mail' but it was only a matter of time.

All of these items required care and maintenance from those whose lives they helped to run. The straightforward demands made by earlier artefacts – such as socks that required darning or a wood stove that had to be cleaned or a fence weatherproofed or a floor needing a mop and bucket – were succeeded by demands that were far from straightforward. He had understood less and less of the equipment that surrounded him: the triple darkness of the three-pin plughole symbolized his limited understanding of domestic technology. The repair man came to the house; or the machine was sent to a place where expertise was centralized. At any rate, RT had spent increasing time 'keeping an eye' on his machines or keeping an eye on the devices that were keeping an eye on them. Breakdown could be an organizational and financial challenge, illustrating the extent to which possessions might possess those who possess them.

The boiler-suited assistants summoned to his house when gadgets failed were also a manifestation of the logic whereby labour-saving devices simply re-defined labour. There was another logic of possessions; namely that they required other possessions to support them. Artefacts needed auxiliary or complementary artefacts, and devices formed teams. This was evident at the homeliest level. Chairs required cushions (buttock-felicity promoters), socks required needles and threads for repair and these (in his childhood) required mushrooms. Plates that replaced the palm of the hand necessitated a plate rack to store them and a variety of cleansing devices, ranging from dishcloths, through brushes of various sizes and shapes, to washing-up machines. Tables needed tablecloths, windows required curtains and blinds and locks. Pen, ink and paper, created a need for blotting paper. Coats demanded coat hangers and clothes brushes, and wardrobes and cloakrooms. Shoes requisitioned polish, hard and soft brushes, cloths to apply and cloths to rub off, shoehorns, insoles, racks to parade them, not to speak of laces (that have to match) and plastic-sleeved ends to permit the laces an easy passage through the relevant orifice. Shirts required cufflinks which needed their own housing to prevent pairs getting separated, and the boxes a designated place so they would not be lost. (And that principle of 'Why walk when you can dance?' was evident in the monogrammed heads of the links.)

Clothes had their own clothes: ties in cellophane sleeves, suits be-suited, and trousers in presses, shirts in their own garderobe. Doors to keep people out had keyholes and keys to let them in, keys required key-holders in the kitchen and key rings in the pocket, and keyholes were sometimes covered to prevent gazes piercing them and dirt blocking them. Knockers and doorbells enabled the legitimate inmates to be alerted to the desire of others to enter. More esoterically, electronic devices required chargers and, for those who travelled abroad, adaptors to fit foreign plug holes. And when it came to cars and boats – well, the list was endless.

And so his inanimate servants were gathered up into networks, and into networks of networks, ensembles, addressing aspects of a single need, connected needs, needs created by the satisfaction of needs, corresponding to networks of ideas, of concepts, and of little worlds. The need to look after those items that looked after him – out of respect for their practical use of course but also for the labour and skill that had gone into making them – meant that the series had no logical or natural end. The things charged with looking after things demanded to be looked after in turn. The idea of a brush charged with cleaning out the brush cupboard was not entirely fanciful. One way of ensuring the welfare of his kit was to store items in designated places. They were not only protected but also easy to locate when, at any time during their 24/7 on standby, they were scrambled for action. Purses, handbags, plastic bags (labelled as required), umbrella stands, briefcases, trunks, cardboard boxes, box rooms, and sheds all played their part in ensuring that RT's insentient servants were warm, dry, sheltered and truly handy when they were needed. And the notion of the 'handy' was a reminder of the Ovidean metamorphoses of the palm from purses, to bags, to carts.

His possessions were rarely simple. They were things of many parts, a multitude of components. The most ubiquitous and basic expression of complexity was the already mentioned handle. His house was a place of handles – and their descendants push buttons, knobs, and switches. The handle, integral to cups, cutlery, garden tools, buckets, flush toilets, bags and cases, doors, acted as as go-between linking the tool-user and the tool. They were a constant reminder to his inconstantly philosophical mind of how much of his living consisted in handling the world by handling handles; a reminder in its turn of the extent to which his agency operated through intermediaries, to the point where some philosophers were inclined to think that the distinctive characteristics of those events that were identified as causes lay in their capability to be used as handles. The verb and noun 'handle' sat snugly united in the palm of his curl-fingered hand.

The range of materials (caged in tubes and jars and rolls) for mandating spatial cohesion between things that were more naturally distinct and apart, that had an inclination to go their separate ways under The Principle of Chaotic Tendencies, was impressive. His dwelling had been richly provided with outsourced adhesive effort, proxies keeping a grip on things even while he slept or was (say) in London or on holiday. Glue, cellotape, masking tape, string, Blu-tack, elastic bands, Polyfilla, nails (shiny new and rusty old), screws, tacks, safety pins, drawing pins affixing notices to noticeboard to ensure that they should be noticed, treasury tags (threaded through prepared perforations anticipating both separations and needful connections), clips and pegs, had accumulated over years. Though their role in his life had diminished as ink on paper was partly displaced by 'information' on magnetic media, they remained indispensable. Some had had an importance disproportionate to their size. In the early part of his adult life, he had owed much order to paper clips, staples, 'bulldog' (respecting their tenacity) clips, and pegs. Some mechanical grips assumed a multitude of roles. The pegs, for example, that attached his wet clothes to the washing line shared a common principle with, and a family resemblance to, those that united papers governed by a common theme, others that sealed the muesli packet to keep the contents fresh, yet others that restrained a paper tablecloth in a breezy taverna by the sea, or were mobilized (metaphorically) to protect his nose against noxious odours, or those that had clipped his umbilical cord to finalize his entrance into the world as a distinct individual, underlining that there was no return to the uterine Utopia.

The ways of interliths (threads, nails and glues) that brought artefacts or their components into needful association and maintained the integrity of that association were thus many and various. Hardly less impressive were the lubricants, the oils, the powders, the creams, and the jellies that soothed and smoothed the passage of one surface over another, reminders of how much he spent on consumer goods designed to adjust coefficients of friction. He used purpose-made slime dispensed from tubes for unsticking the stuck and soap for dealing with the sticky (the latter prompting more fuss than it justified, perhaps because it was a primordial source of discomfort that carried much symbolic weight); and he relied on deliberately abraded surfaces to guard against uncontrolled slips, so that controlled movements were possible even when he walked over ground rendered 'treacherous' (the usual personifying metaphor) by ice.

His ideas of domestic order had sometimes been quite liberal. Of course jackets should be hung up in the wardrobe on coat hangers but it was usually enough to drape them over a chair, where their contents (and for him a jacket was often a prêt-à-porter office) would be handy or (in a public place) as a marker of a seat that had been bagged. This was a common example of the multitasking that was often imposed on his possessions. The door was fitted handily with hooks so that it could assume an avocation as a coat rack, where he could store outer layers not required once the wearer was inside the door, itself a marker of the line between the weather without and the calm within. Handkerchiefs purchased to harvest the effluent of the nose would double as a blossom in a jacket pocket, and for some, knotted at four corners, a device to protect a naked cranium from solar radiation (for him a memory of his 1950s childhood, preserved on a postcard), or in the case of farewells, both magnify the waving off and wipe away the tears prompted by the resulting absence, or the idea of it. Mugs held down papers when there is the threat of displacement by a breeze; and coasters, designed to prevent cups leaving marks, were also deployed to conceal those same marks while acting as a proxy memory of a summer holiday.

And his every possession was capable, he had noticed with respect to his fellows, of signalling something about its possessor. This was most elaborately developed in the case of clothes: epaulettes and medals, T-shirts that refer to themselves, a conscious crumpledness, made the wearer legible and in many cases guided the reader as to the interpretation he should make. (The modes of studied dishevelment, the appearance of neglecting one's appearance to broadcast that one was attending to other, more important, matters, were many and various.) And at times of extreme need, no goods with the desirable material properties were excused call-up from duty as a weapon. Of which more presently.

Most hobbies had left him but he had once collected stamps and his children had collected stickers, adding to the pile in the loft. Neither party dreamed any longer of completing collections but there had been a time when many books passed through his hands as he pursued the mirage of omniscience – or at least of having read, if not remembered, everything that, according to the strictest criteria, was worth reading. He had 'left behind him' (a standard phrase suggesting that he was showing his bottom to the world that should cause a little jolt) many involuntary collections – back copies and odd socks being just two examples.

The artefacts and utensils – the inanimate servants of his needs, appetites, and duties – which had supported him were not all stored at his address. Nor, in many cases, was his ownership exclusive. He was the beneficiary of shared amenities to whose creation and maintenance he contributed in accordance with rules and statutes of citizenship which, if spelt out, would have occupied many thousands of pages. Pavements, roads, lamp standards, power cables, the infrastructure of the utilities that called in at his house like any other, public buildings (including those where he had treated his patients and the jealously guarded real estate where he had his department and his office), parks with their furniture, were some of the most obvious of these shared possessions, owned by no one in particular because they were owned by everyone in general, or at least everyone who had the right to visit and remain.

The address he called 'my address' was located in a town which he referred to as 'my town', itself located, of course, in 'my country'. The 'my' in each case corresponded to a different kind and strength of possession. The upshot of this was that the greater part of his life was spent in an artefactscape of items, ranging from pins to highways, from gravy boats to laboratories, which filled every conscious field, forming an enclosing theatre that shaped and supported the life he was shaping for himself.

Though, as the years stacked up, his hoard of possessions grew, like his CV, and, to some extent his belly, by the time he had reached old age most things he had ever possessed had simply passed through his life and out of the other side, destination unknown or at least untracked.

This was most literally true of 'consumables'. Food and drink came into his house, proceeded to his mouth, yielded its hedonic potential and delivered its metabolic payload, left his body by the designated sphincter, and exited his address through the sewage system attached to the appropriate facility. There were exceptions; for example, jars of condiments that stayed for years and wine laid down awaiting a change of adjectives, but transience was more typical. There were possessions that had a portfolio career, either as hand-me-downs that worked for a succession of owners, or through acquiring a retirement avocation. Not infrequently the latter involved something of a cut in status as in the case of an expensive silk blouse that passed its dotage as a duster; or a flamboyant tie used to tie up a matress to be taken to the dump; or a discarded toothbrush that proved ideal for extracting dog dirt between the treads in the sole of boots. Such examples made him glad that there was such a thing as time that prevented everything from happening at once. And they pointed a lesson as to his own future. Between short-lived consumables and consumer durables such as furniture that had something like tenure were items that stayed as long as they remained intact, in fashion, and were still needed. Cups that cracked, trousers that had once been fashionable and now looked like the idea of something that someone desperately unfashionable had considered fashionable, and cricket bats that were no longer held by a player dreaming of the stroke that would send a ball into outer space, found their exit visas stamped en route to the council tip – or the charity shop, where one man's rubbish could turn out to be precisely what another man was looking for.

The level of damage at which items qualified as being beyond repair, and an injury deemed fatal, fell steadily during his life. The instinct to 'make do and mend' weakened as he (and the nation) grew more prosperous. Such, as a consequence, was the flux of goods that his habitat had sometimes seemed like a mere conduit. To take just one example, a trickle of cheap reading glasses – to wipe the mist of '–ish' off the printed world – entered his life in his mid-forties. By his fifties, this had quickened to a brisk current, checked only a little by new methods he had devised to anticipate and pre-empt his absent-mindedness. The extent to which the lost exceeded the found seemed almost to amount to a breach of the law of the conservation of matter.

Yes, the list of the permanently lost, the mothballed, and the dumped, was dwarfed by the register of the mislaid. A considerable portion of his later life was spent in looking for items just put down; just 'a second ago!' he would cry, to underline the discontinuity. Pens that slipped under papers or rolled silently to the floor as if determined to extend their useful life by protecting their store of ink were chief among many such items that appeared to have a Berkeleian tendency to go out of existence unless continually watched. The already referred-to examples of spectacles were able to hide in the plain sight their absence denied him. They could be on his forehead as, with shortsight not lengthened by their assistance, he would search for them. This illustrated, by default, the more usual miracle of the unity of the conscious moment of an embodied subject who was able to look at one item, think about another, remember a third, and still (except on occasions when his glasses were involved) be aware of the distribution of small objects around him.

And smallness was an issue. On the many occasions when he mislaid his car keys, he was haunted by the knowledge that the item in question was so small compared with the place in which he was searching for it – a room, a house, a street, a city, a world. It seemed astonishing that the mislaid was ever found – particularly as the signal of the desired object was lost under the noise of other possessions. This prompted an intermittent obsessive tidiness and the ingenuity, already referred to, with which he attempted to forestall the consequences of his own absent-mindedness. The mildly exasperated cry 'I had this in my hand just a moment ago!' was the surface seepage of a deeper anguish at his own discontinuity.

Natural wastage notwithstanding, there remained a growing heap of possessions, of things not weeded out by obsolescence, mislaid by accident, or swept away by changes in his life or an incoming tide of new acquisitions. Ornaments, which by definition had no clear function except to look beautiful or interesting or impressive, were particularly tenacious: having no use in the first place, they could not be sacked on the grounds of uselessness. And the longer they stayed, the stronger their case for staying on: their aesthetic rationale was supplemented or supplanted by their curatorial function. They came to commemorate a place or a time from which they had been bought, and which had itself become more precious for dwindling into ever greater distance. Chucking out a souvenir colluded with amnesia and exterminated a little bit of himself. There were many things with which he had no emotional bonds that were retained for theoretical reasons. A fish-slicer, for example that had divided no fish for several decades. But then wasn't this one of the wedding gifts? So the cloth that had covered his first married breakfast table, the single survivor of a pair of earrings bought to celebrate the birth of a child, stayed on along with a cheap wooden elephant bought from a street trader in Nice one hot summer evening because it was a reminder of Nice and a hot summer evening in which a trader sold him a wooden elephant.

There were accidental souvenirs – hidden underneath a bed, behind shelves or radiators, or beneath cushions – that, like sheep's wool on a blackthorn tree, had snagged on the edges of the tangled undergrowth lining the passage from entrance to exit of his lifespace. Tickets were particularly evocative: the Metro *billet* found in the pocket of a coat not worn since the previous winter; the ticket that bought them a couple of hours of a June evening in the car park near the sea years before, located under the rubber mat beneath the driving seat; the unexpected find in the wallet of a ticket to a performance of Mahler's Ninth of five years ago, of which he remembers only that it was 'unforgettable'. A change of clothes to mark the seasons often turned up particularly poignant memorabilia. In his childhood, a jujube stuck to a farthing in a winter coat mothballed for the summer season. In his adult life, a pang-bearing programme, perhaps – for a concert from last spring, when the evenings were getting longer. It invited him to look through the intervening months, with their floral and arboreal markers and the events that had unpacked themselves from appointments in the diary to full-blown mornings and afternoons and evenings, and prompted him to remember birdsong from bird silence, and glimpse the depth of his life. Rather as the first light wrapping of condensation on the car had looked back to the frost last seen in April and January's snows.

Hardly surprising then that, like many others, he was a case of the Diogenes syndrome, though his was not severe. It was a way of respecting past selves, avoiding a shallow 'presentism' in which this evening's concert would eclipse all the concerts and his life would be a succession of 'ands' or 'and on to the next thing'. The past was most literally present in the microtome-thin samples of pickled light that were present in every room of the house. The living and the dead, past bodies and selves, past places, offered themselves in moments of attention snatched from the permanent hurry or restlessness or fidget of his life. Courtesy of the misprinted mite – his eighteen-month-old self running on the beach towards the camera and an exasperated father wanting him to keep still – of his children in academic dress on degree day, of his mother as a little girl, and as a widow of ten years, he lived side by side with many layers of lost time. If their remoteness had not given him sufficient pause for thought it was in part because they were there every day, naturalized in the present, in part because their frames seemed to enclose them in brackets, and because they were always to one side of a (more important) future that was making him hurry past them. They were also overlaid with contemporary reflections painted on the glass that covered the ancient light: slicks of brightness from today's Christmas tree, moving shadows of thick newly opened foliage, and images of the faces looking in at the faces that only seemed to look out.

These souvenir portraits were supplemented by other, quite numerous, somewhat random, samples of outside: an aerial photograph of the house itself, snaps of a familiar beach and a familiar mountain, and (in the lavatory) a group of men lunching on a steel girder near the top of a New York skyscraper. The jumble of images – the French poet Baudelaire looking sour, Port Isaac in pencil, Venice half-drowned, a map of Europe – was as crazily varied as it was numerous. The densely populated walls, mantelpieces, window-sills, and display cupboards added to the (correct) impression that his dwelling was almost like a silkworm's cocoon, though its substance was not so much his own flesh as worlds he had made his own by living in them.

There was another reason for the quickening pace of consumption and accumulation: the growth of the GDP attached to his caput. Money was the key meta-possession that made possession possible, the general possibility of owning or consuming. His trajectory from childhood to late middle age had been shadowed by an increase in an indirect, and yet desperately concrete, form of power: purchasing power. The boy looking for pennies in the garden, or behind the sofa, the grateful recipient of pocket money, had passed through grants and bursaries to a salary. The monthly stipend grew year on year: with seniority came pay grades and pay rises and eventually the bank statement ceased to be a menstrual tension. His relationship with money, at first defined by coins jingled in the pockets of his short trousers or proffered in a sweet shop, or collected in an incontinent piggy bank, became increasingly abstract, taking the form of numbers on a pay slip, a balance sheet, eaten away by cheques, standing orders, repayments, annual statements, or various devices requesting his Personal Identity Number. The transition from the shiny half-crown to weightless numbers, to abstractions whose secondary qualities were irrelevant, was a metaphor of what was happening in his life: more meaning less, attenuation of experience, increasing autonomy, progression to a kind of weightlessness before gravity asserted its dominion.

The consequences, built up during a long life, were entirely predictable. He reigned over an empire gained, like another more famous one, largely in a fit of absence of mind. Admittedly, the sheer scale of his ownership was to some extent hidden from him because much of what he owned was 'tidied' away. It was a false tidiness. In many cases, it was one that had mocked the grammar of possessions, replacing it with the law of the jumble in which raw 'and' reigned unchallenged – concealed of course behind the smug, bureaucratic geometry of the cupboard door, with its understated ornamentation. The rational order exemplified in the set table – where knife, fork and spoon were arranged with the plate and the wine glass and the place mat to form a setting joined with other place settings around the central candles or flowers – or the hierarchical arrangement of occasional tables – was lost. The halfway station to chaos was marked by the neat stack of similar items, standing to habitat as the lexicon to poetry. Eventually, this crumbled to the kind of disorder, an arrested pell-mell, in which teddy bears would lie down with gas bills, a broken thermometer with a crayoning book, and higgle with piggle.

He had deployed all sorts of strategies to keep above the encroachments of disorder and the impending triumph of naked 'and': labelled drawers, racks where plates and CDs stood to attention, toolboxes with their structural logic, medicine chests separating tablets from lotions, and so on. But sooner or later the disorder of the pocket or the handbag would be replicated in expanded form in the storeroom, the attic, the basement – and the loft.

Ah, the loft! A limbo where there gathered items that could not be thrown away but could not fully justify their retention. It was the land of 'maybe' ('we might need this one day') and it gradually filled with an alluvium deposited by the currents of life that had passed through the house. He had regretted this costiveness when he had moved out of the house he had occupied for the years that had encompassed most of his children's childhood and early adult life.

Though it had been entirely voluntary – he had not been evicted by poverty or ill health – the house move had been nonetheless disturbing: it forced him to realize that this most longstanding of the fixtures of his life had been a stage set. The appearance of permanence was unmasked as pure Potemkin, as if concrete proved a smoke of dust. The bared walls, naked echoing floorboards, unfurnished spaces – that gave a chilly, churchly timbre to the voices of the outgoing occupants, as they had bade goodbye to each room in turn, and to decades of shared life – said 'Not only will we outlast you but every mark of you will be effaced as you effaced my previous owners. You will be painted, decorated, rebuilt, and gardened out of the picture. A new flock of birds will nest here and be as at home as you have been before they in their turn fly on.' The house's future, void of his presence, was a premonition of a world carrying on oblivious of his absence. He had had a foretaste of this continuation without him on returning from holiday when he was met in his study by a rusting half-eaten apple and thick-skinned half-drunk coffee, in the kitchen by newspapers that had grown dry and dusty, and in the garden by flowers that had bloomed and faded – in short by a place that had got on with its own material existence without him.

The immensely complex logistical labour of the house move, arising out of the quantity of his possessions, had damped elegiac reflection. As one house was emptied, and another filled up, his possessions were reduced to their most basic material qualities of weight, size, shape and number. The grinding repetition of packing, folding, lifting up, transferring, and loading made RT feel that, far from adding to his existential heft, his possessions had squeezed him to the edge of the space he had owned. He sometimes thought with affection of his long-deceased dog (another possession that had passed through his life) whose touchingly small inventory of effects hardly extended beyond a rubber bone, a much-abused teddy bear, a basket and blanket, an intermittent idea of territory, and a bladder full of signals (with minutes of credit measured out in squirts of urine). The dog's unencumbered journey through the world was an involuntary sermon on the art of doing without, or with less, or at least of being content with what one has before it starts to pile over one's head. Even books proved ultimately to be more like material objects than ethereal clouds of meaning. This had already been highlighted in the most literal possible fashion when his mother had used the signed copies of his own (often rather heavy) works to stabilize her shopping trolley. Or when they could be exploited for their obstance, acting as stops that kept doors open against a wind that threatened to slam them shut.

During the removal from House A to House B, as packing case after packing case had been packed, pack-horsed, and unpacked, the back-to-basics nature of the labour had prompted the radical, and bleak, reflection that the many protective layers wrapped round him by possessions were equally subject to the laws of physical nature as the things from which he sought protection. None of this clobber could rescind the conditions that governed the natural world. Not only do precious objects rust, rot or break – the egg cup falling under the same dispensation as the egg – but they may break us. Those who live by the spade may die by it. The brick intended to shelter us from harm may drop on our foot, the wire bought to exclude trespassers may trip us up. Whereas he did not subscribe to Adolf Loos's claim that 'ornament is crime', RT had to concede that ornaments could be used as criminal weapons. Mementos hurled with sufficient force at his head had the capacity to blot out memory. Toppling masonry, bearing the imprint of the classical past, could plunge the present into darkness. Roads, or the traffic on them, may be as dangerous as rivers. Self-propelled modes of transport may career out of control and horsepower be transformed into the motor of a tiger on whose back he was riding. A bus taking children safely to school may pack a punch indistinguishable to the recipient from that of a boulder rolling down a hill. A loose carpet, designed to soften the hard surfaces of a tough planet, could deliver the fatal blows from which so much of his life was devoted to insulating him.

Indeed, home was set with traps that had the power to cause the potentially fatal slipping and tripping of upright man, a fall that would be the first step in a descent from humanity to materiality. It was capable of dealing lesser wounds: the scars of bangs, knocks and stubs testified to the melancholy fact that the third law of motion – action and reaction are equal and opposite – was not suspended inside his retreat. Drawing pins were particularly vicious thorns to the naked foot, knives edited the meat that held them, buzz saws be-handed. Electrical appliances electrocuted, gas fires gassed, and baths and swimming pools drowned those who had installed them for warmth, light, cleansing and recreation. Thus the revenge of objects, expressing their boiling resentment at a lifetime of subservience. Purchased to ensure health, safety, comfort and fun, his possessions had had the power to bring illness, danger, discomfort, and dismay. The bitter irony of the fatal domestic accident was overlooked perhaps because it was so commonplace, though it shared something of the particularly sickening nature of domestic violence: suffering and danger in a place that should be a refuge from the world, a place of peace and safety.

Yes, artefacts had insulated humans against heat and cold and thirst and hunger and predators and accidents. Yes, they had transformed what filled a human life: consulting a bus timetable replaced trudging over the sticky earth, or poring over the contract for the hire of a mechanical digger spared the user delving into the same recalcitrant earth. Yes, they had created new preoccupations unknown to biology – such as planning the stamp collection. And, yes, they had enriched the seasons with lights and shades unknown to the forest floor or the immemorial mountains. The impossible had been brought down to earth, as fairy lights and mobile phones and computers delivered what fairies, demi-urges, and gods could not have offered in response to the most urgent and propitious of prayers. But the laws of nature were not to be suspended, either for him or his possessions. How could they? For the spade could not have assisted the necessary (if arm's-length) engagement with the earth without falling under the same jurisdiction as said earth, as revealed in the contribution it made to the wearisome weight of the spadefuls. Hence the abusage that was the dark side of usage. On stage and off, the door's acoustic properties were exploited to express the anger of the person passing through it. It was a primitive shout that few mouths could match and it offered no opportunity for reply. Glass had evil powers as the faces of those who had been 'glassed' testified. Broken glass seasoned walls with an acidulated edge designed to harm trespassers who scaled them. An innocent walking stick could be transformed into a cane and a baseball bat was a hideously effective way of beating the life out of a fellow human being.

While abusage was a kind of betrayal that depended on using our possessions to collude with nature and the uninflected laws of physics against our fellows, it had its lighter side, as exemplified by the idea of a saint using a halo as a Frisbee or a missile that might decapitate a fellow saint. There was a kind of glee in the sound of smashed glass (this licensed vandalism being an incidental reward of RT's trips to the council recycling tip), in tables crumpling so obligingly in those bar-room fights in traditional Westerns, in pews used as kindling, or tower blocks under a demolition order, crumbling like dunked biscuits after a splinter-second pause of disbelief, before the life spaces, the snuggeries and cosy places, of thousands were reduced to rubble, dust, and smoke. The idea (a favourite of his boyhood comics) of a precious volume (let it be the Bible) stuffed down the seat of the trousers of a miscreant, to intervene between the cane that punished and the bottom – the traditional seat of castigation – that was being punished, delivered a reliable quantum of subversion. And the use of the *Daily Mail* to absterge the podex when the toilet roll had been slimmed to the central cylinder of cardboard had seemed to him an appropriate critical response.

So his possessions – ingenious, innumerable, tireless servants of his basic and elaborated needs, of his comfort and joy, mitigating drudgery and occasionally affording delight – could not annul the material conditions that, since they had brought him into being, had also dictated his prior and subsequent non-existence: the processes that had generated him and maintained him from day to day were the same as those that guaranteed his disintegration. While the artefactscape that enclosed him improved his odds of continuing in a controlled, intermittently pleasurable life, and could protect him against the 1:10, 1:100, 1:10,000 disaster, they could not reduce the infinite odds against the existence of the singular creature RT. The smugness that sometimes came with a snugness that silenced the menace and sorrows of Outsides was sooner or later trashed.

And the predictable result was clear: space that he had made his own – to the eyes of theft and envy a comb honeyed to the brim in every cell – set in greater spaces that belonged to others, to everyone, and to no one, was now consequently vacant of his possessions and the orphaned agents of his agency were no longer executing his wishes. With their assistance he had created a stage where he had been able to invent or discover the stories of his life, and avert his gaze from the Big Story that had only one ending. But now that ending had arrived and he had come to this: between a mattress and a sheet, neither of which belonged to him, dispossessed of all possession, all 'having' now had, imperfect now pluperfect, a naked ownerless body, he had ended as he had begun.

eleven

Semantic Space

Visitors paying their last respects will direct them to the capital of RT's body, to the head with which they had been tête-à-tête for so long. It was here, more than any other part of his frame, that revealed his changing take on a changing world. Little of its meaning-packed anterior surface had been excused the duty to communicate: mouth, eyes, nose, forehead, cheeks, all had their say.

His eyes had had a richer repertoire than his nose. He had flared his nostrils, but he had wrinkled them to gather up a wince, express doubt or report a dubious smell more rarely than many others he knew. When he tilted his head to turn up his nose at what had been brought to his attention, it was usually in inverted commas. And, as noted, he had never snorted with disbelief. Like the forehead, with its frowns and its 'looks' of surprise, his nose had not communicated in isolation.

The gaze could, when required, convey anger and hatred with a fixed stare set in a dilated orbicular orifice. Many modes of sadness, ranging from grief to reproach, had been signalled by the moistening of the cornea. A momentary unilateral curtaining of vision had signified complicity, support, or shared knowingness. Rolled up, his eyeballs had emitted the visual equivalent of a sigh at a competitive dinner guest showing off, the pretensions of some contemporary art, an unexpectedly large gas bill, the Chancellor of the Exchequer's tax plans, or the incorrigible nature of humanity. What power it had to convey so much critique, solidarity, or ironic distance, or to pantomime despair, or superiority, with so little!

Even so, it was his mouth, the expressive epicentre of the face, that had been the richest fountain of signals. Its carnal repertoire was impressive: the smile (ranging from empty emoticons to dazzlers double-bracketed by the folds of orbicularis oris at full stretch), the omicron of surprise (that may or may not have recruited italicizing whistles and even shakes of the head), or the omega of gape-mouthed, real or pretended, amazement, not to speak of grimaces, winces, looks of disgust, of arch disbelief, or even mockery, grins of suppressed laughter, lit with knowingness, perhaps even inflected by a poked-out tongue (taking time off from licking envelopes and spoons, from playing its part in controlling a bolus of food en route to his tissues, or glossal stopping the glossal stopped), that might itself be in inverted commas as if mocking an earlier, playgrounded version of himself in which his poked tongue was a serious weapon, or sucked teeth in mimicry of a garage mechanic's pessimistic expectations of the possibility of curing an ailing jalopy. No wonder he had sometimes passed judgement on his own smiles, and suspected that they looked insincere or foolish, and, as a teenager, had even practised them in the mirror.

And he had smiled out of love, humour, satisfaction, pleasure, greeting, to reassure another – and in illustration of the smiles of others. In a life buzzing with words, his relationship to his own smiles could be very folded indeed, especially when they recruited other parts of his face, such as cheeks blown out in panto surprise or (with the aid of a mimed mopping of the brow) the condition of being overheated, or (with a shake of the head) a certain attitude towards an unsatisfactory, or even 'mad', world, a world he has now lost.

Whatever apparent expression his face will present to those who have come to view him, it will be entirely void of expressive intention. The organism has breathed its last breath and RT has smiled his last smile and spoken his last word. Nevertheless, it is difficult not to see something there – even if it is only the peace we have dismissed as fake – just as when we look at a skeleton with its teeth uncurtained by liquefaction of the surrounding flesh, we may fancy laughter at a world in which it is no longer a player. RT's mouth, where the absence of meant meaning is most evident, may trigger memories of the astonishing virtuosity of this structure. To see this clearly, we need to step back a little from the semiosphere.

Its most basic function (and more broadly that of his oral cavity) – ingestion – was one also performed by the stomata, suckers, and comparable appendages of other living creatures whose path through the world is innocent of all those preoccupations that had made up his ordinary human life. He ate many thousands of meals and snacks for the pleasure of eating as well as the necessity of nutrition and drank many cognate drinks, more often for pleasure than necessity. He took in, chewed, swallowed, spat out, chicken, chocolate, and Chardonnay. His days were structured by food and the kitchen was the heart of his home. The dinner was the high-point of the festivals that were the mandated high-point of the year. Flash back to Christmas Eve and to a queue outside of the butcher's shop – talking clothed, living meat, moving towards piles of silent, naked, dead meat. In pursuit of scoffing, swigging, nibbling, sipping, masticating, and glugging, his mouth recruited his hands, and requisitioned an armoury of instruments. Between hand and mouth there intervened the wherewithal, and the activities necessary to get it together, which purchased and processed whatever the hand brought to the mouth. (And it was not unusual for the mouth and hand to cooperate in anomalous ways as when, accessing his food, he bit through recalcitrant wrappers in the already alluded-to War of The Packaging against the People.)

While he had not been one of those individuals who endlessly chewed the inside of their cheeks, distorting their faces with an involuntary appearance of unilateral wry scepticism, his mouth did engage with itself. He deployed his tongue to prise fragments of food from damp nooks where they might decay (better that than using his finger, something that sickened him in others); to test how well his teeth were brushed, as recommended in an advert for certain brands of toothpaste; to moisten his lips, redistributing saliva made scarce by anxiety, to ensure that those lips did not snag on one another or on his teeth when he spoke in public; or sometimes, for pleasure, to tickle his hard palate. Occasionally he accidentally inflicted exquisite pain on himself – a sensation unchanged from childhood to old age – by biting his tongue, as the components of the mastication machine got tangled. This rare blunder was a lacuna in an otherwise miraculously precise organic self-knowledge. Such knowledge was also utilized in keeping a potentially bitter tablet from dissolution until he had found water to wash it down, or in exploiting the anti-tussive power of saliva to damp an incipient cough that would have betrayed him in a game of hide and seek, would have woken Mrs RT from much-needed sleep, or shattered the post-climactic silence at the end of a symphony.

Some of his oral outputs had been of marginal significance. While the fear of vomiting had been a prominent feature of his childhood (especially when he was trying to get to sleep), the reality had been very intermittent. He spat not infrequently after cleaning his teeth or washing his mouth with antiseptic deodorant. But the contribution his mouth made to his presence in the world had been mediated to a much greater extent by sounding air. By this medium he was able to give out the most, and most varied, pieces of his mind, and to engage most effectively and cost-effectively with the world.

Conveying significance by means of exhaled air was not unique to the species to which, until yesterday, he had belonged. His speech was rooted in a more fundamental capacity to transmit meaning by means of the sculpting of breath, though the passage from the roots of airborne communication to the leaves of utterances woven out of more or less grammatical sentences remained a mystery. Cries, shrieks, of protest and shock and amazement, go straight to the heart of the hearers, but these had not been his usual style. He had emitted a share of groans, ughs, and oops, sometimes woven in with the skulls of dead deities – 'Gawd!' rather than 'God' – but even then he was sparing. The occasional three-note yawn – like an orphaned fragment of music – broadcasting boredom with the moment, the conversation, the topic, the speech, the company, life or the world – had been allowed to escape him. He had not been a great one for sighs – that deep breath in taken to create a deep breath out – though he was aware how much they could convey, and with what economy: a free-floating weariness, a silk scarf on a shoulder shrugging indifference to something in particular, or contrariwise an inexpressible sorrow at someone's demise, or impatience at being kept waiting in a supermarket queue. And he had been aware that there was 'Sigh!' available to him as a speech act that pointed to itself.

Such complications could be unpacked even from those accidental gaseous emissions to which creatures with a gastro-intestinal tract are prone. RT who has now emitted, and laughed at, his last belch, enjoyed eructations for themselves and as a kind of leitmotif in his life. The schoolboy contemporary of whom he remembers nothing else once caused a belch to echo from the pavilion clock. His father, in the years before his dyspepsia was cured by getting false teeth, used to emit spectacularly ornate belches (they seemed to have corners in them) at the meal table. This had prompted another kind of emission of air – laughter – that was tolerated so long as it did not go on too long and of course it always did because the knowledge that it had gone on too long made it more difficult to suppress. (The solemnity of his corpse makes it difficult to grasp how often he had been, or accused of being, or felt, silly and been instructed to 'wipe that smile off his face' – a command now obeyed forever.) And finally, he remembered – long after she had died – his mother remarking with a rare cultural relativism that belching was regarded in the Middle East (a place associated in his childhood mind with a few Orientalist images) as a polite signal of food enjoyed. Thus was his ten-year-old self kept afloat in later time-slices of RT: on exhaled air, exquisitely carved by the passages through which it had exited. Many years later, his children had replicated at regular intervals 'Bloody Hell Baldrick!' (Don't ask!) in speech fashioned out of belches. And this trans-generational solidarity gives an entrée to a reflection on the use of 'follow through' – by means of a metaphor drawn from the game of cricket – to refer to the unfortunate consequence of enthusiastic release of wind.

A cue perhaps for an exploration of the various modes of laughter that rocked him – guffawing, chuckling, chortling, giggling, and sniggering, ho-ho-ing and tee-heeing (all without the slightest hint of eldritch) – or which he simulated out of solidarity, politeness, sycophancy (where he made the dutiful seem uncontrolled) or out of fear of being marginalized as the one who did not get the joke. Or of grunting, harrumphing, humming, hmming, hemming and harring, brrr-ing and aagh-ing. We have, however, postponed too long our engagement with the millions of words that that silent mouth had spoken, with a lifetime of enunciation terminated in the stillness of those lips, whose only movement now is an imperceptible stiffening prior to liquefaction. We therefore pass over instances of raspberry blowing, whistling (he was the frequently whistling son of a frequently whistling father), singing, yodelling, or (on a protest march) of bazooka playing, to come to the most mysterious of the modes of intelligence borne on the air – different from the sound of rock falling on rock, the chuckling of a spring, the rustling of leaves and grass, the tread of an approaching predator, the songs of avians, the trumpeting of elephants, and the barking of dogs speech.

The challenge is to map the scope of his speech, while recovering the worlds sidelined by the horizontal haste of utterances journeying to the end of their sentences so that requests may be lodged, objections raised, explanations offered, chases cut to and stories told. Trying to do justice to a lifetime of speech in a few pages is rather like using a few drops to exhibit the ocean, or sounding its depths in a bathysphere made of saline. It reflects the wider difficulty of capturing his life, or life itself, in a nutshell. How can this Wednesday afternoon stand outside his life to see it in a totality that includes (as a small part) this Wednesday afternoon? How likewise can a few paragraphs' worth of sentences rise above language to survey the totality of his frontals and asides, pronouncements and mutterings, including the possibility of those paragraphs or sentences? To break the impasse let us step back a little and situate language in something even wider than it, albeit using scraps of language. Let us recap.

His senses opened him up to a space in which he was located. This space was expanded by being a shared space, most obviously of the eyes and ears. Others' eyes and ears might look and listen into what was for him a rumour of places beyond the places he knew first hand. Words that came from others – incorporated into a common tongue long before his arrival at himself – increased a millionfold the number and kind of spaces in which he lived. Mummy and Daddy, and siblings, and teachers and mentors, and books and newspapers, added Mrs Smith, the playground, London, Africa, the North Pole, and Mars, to the cot, bedroom, staircase, kitchen, cupboard, spaces in which he peered, crawled, walked, and ran. His own past, and the past in the keeping of his species – the Battle of Hastings, and the origin of humanity, life, and the universe – supplemented whatever at any present moment he had directly recalled. Consciousness shared through words extended 'Out There' – beyond what he could see, hear, touch, taste, or smell – into something he came to know or merely 'knew of'. He was aware in consequence that his awareness was confined to a small part of the world. Whispered asides, lists of facts, encryption, and technical terms, reinforced his intuition of wheres and whens beyond his here and now or his here and there. For most of his life, he had effortlessly outsized himself: that now wordless mouth had said 'World', 'Universe' or 'Africa', 'Welsh history', 'Carbon'. And this outsized self was present even in the voice that said 'summer evenings'. The verbally mediated notion of evening summed an unbounded series of evenings; and even the idea of one evening encompassed countless people, streets, cars, clouds, trees, houses, conversations, glimmers on fruit bowls, children trying to get to sleep, and glances exchanged between strangers, friends, and intimates.

His words, then, had gathered up space in ways unknown to the eye, the ear, and the hand. He had spoken of 'London' and 'economic trends'. These vast territories – boundless not merely because ill defined – were cut out by the linguistic tokens shaped by the cooperative activity of his tongue, lips and teeth to bite-sized fragments of a hypersurface, scissored to inwardly digestible items no carnal mouth could have chewed. Effortlessly, he had united the disparate – evoking symphonies, cathedrals and the electric grid in the same sentence as examples of human achievement. With equal ease, he had divided the one into many, separating a stone from its weight, colour, its location, and its use, or a smile from the face that had worn it.

The spaces into and of which he had spoken had more dimensions than a string theorist could have dreamed up in the most desperate attempt to shore up a crumbling theory of the sub-atomic world. In the realms opened up by his lips and those of his co-fabulators doubts cast shadows that can be denied to affirm certainty and truth is a material whose smallest unit is the iota. Rhymes can be lifted from the words that carry them, sneers from faces, and 'not' has as much presence as that which it negates.

The alchemy of RT's mouth was also able to translate space into time or space-time into pure time: in common with his erstwhile companions on earth, he had had a robust sense of the Trinity, of the beginning, the middle, and the end that shaped the many stories he told and audited, including the master story of his own life. Spoken, these stories in turn were located in space, being expressed in sounds that originated from his mouth and faded at the edge of an imaginary sphere defined by the ears that might harvest them. This added layer, elaborated when he spoke out of the side of his mouth to a companion walking next to him, or as tales were being told as he closed in on his interlocutor, or when he was (say) climbing a staircase linking the place where the story happened to the place where the story was delivered, miraculously did not interfere with the time sequence of the locutions in which the story was wrapped or the space-time sequences narrated in the unfolding story.

None of this does justice to the worlds opened by his and others' speaking mouths. A few 'triangulations' – in the geodesic sense – might give a faint indication of their rich variety. If these instances seem terribly particular and rather mundane, this is no accident. To scope the ocean it is not enough to engage in a quantitative story of cubic leagues; for this would overlook the waves (the swell, the breakers, the terror of the eagre poised over the ship), the breezes and currents, the fish, the islands, the sailors, the swimmers and the drowned, the hopes and disappointments, paid and unpaid wages, joys and tragedies.

Let us begin, therefore, with the humble greeting, and his carefully calibrated (if misleading) sense of whom to greet, when, with what greeting; when to be formal, informal, vernacular, or imitative and parodic ('Hi, dudes', 'Hi, grooves', 'Bless you, children'). His echoes, as when fifty years later, he repeated the ironic *Salut du matin* of the gay charge nurse in the ward (where, as a medical student, he was being treated for pneumonia) to an alcoholic patient with pneumonia who had kept the ward awake chasing and being chased by demons, with, 'Good Morning, My Lord Ross, slept well I trust?'

Less astonishing to the world and more astonishing to himself – because it was the general tendency of maturing humans – was his increasing fluency over the years; in particular his ability to seem to know what to say (or to have something to say) after the initial greetings. In part, this was due to his willingness to borrow fillers from others, and even modes of leaning forward interestedly to someone who was talking, to convey the impression of respectful, even undivided, attention. And then there was farewell, which he had always found more difficult. A Keatsian awkwardness affected his adieux so that life after the colloquy seemed to be on the far side of a high wall to be climbed. It took a long time to master the art of saying a goodbye that had the right shape, was delivered at the right time in the right manner. He dreaded its being spoiled by an accidental meeting immediately after a splendidly executed separation. (He had known that he was not alone in dodging an immediate encounter after both parties have agreed with sincere warmth that 'We *must* meet again soon.')

Greetings and (to a lesser extent) farewells were the paradigm speech acts. He had *done* things with words, deliberately and inadvertently. There were many thousands of verbal doings in between Hellos and Goodbyes. Although all utterances were speech acts, some wore the nature of their acts on their sleeve. Saying thank you, for example. He was a thanker – and enjoyed being such – and was easily hurt by individuals who didn't bother to thank. (He had often noted that the paper bags in which his purchases were placed, with their thanks in large print, were sometimes more courteous than those who had served him.) And apologies. He had given, and received, many apologies in his life – given more easily and received with more good grace as he got older. This trend was in part because the notion of pride based on current rather than deposit account had seemed to matter less. So he no longer felt the necessity, when saying sorry, to do so in a non-sorry tone of voice.

This may also have been because the public spaces resounded with apologies. There was hardly an event in history that a politician had not apologized for in order to conquer a little of the moral high ground at negligible moral cost. And the air was resonant with apologies addressed by an official voice, computer-simulated, to no one in particular, to everyone in general, when trains were late or cancelled. This was taken to a thought-provoking extreme in the case of an apology on the forehead of a bus: SORRY NOT IN SERVICE. It was perfect because the words were located just where disappointed passengers at a bus stop would be looking to identify the number and destination of the bus and the itinerant apology was translocated down the route a bus in service with nothing to apologize for would have taken. In short, the information was provided at the place and time of need. Indeed, it was not unfair to deem that the apology sought out, even created, its own consumers. And all of this was ordinary, everyday, language of the kind he had heard himself speaking hour after hour, day after day, year in and year out.

The carefully gauged quantum of aggression, warmth, anger, humour, affection, sadness, protest, joy, and kindness conveyed through the tonal envelope and a dozen other regulated parameters of his sentences was a testament to an extensive cognitive hinterland behind the least considered of his utterances. Various histories (of the world, of his nation, of its language and of the invisible choirs of his fore-speakers) had shaped this hinterland as had his own cultural background, class, family dynamics, schooling, professional colleagues and other groups, not to speak of his native wit, and the way he had used that wit to ease his way through the world and advance the cause of RT.

So many influences had been brought to bear in defining the distinctive dialect or oligolect, colouring the meanings gassed out of his head. His vocabulary, the phrases he had used, the structure of his sentences, the tones of voice and the tonal envelope that characterized the stretches of that voice, belonged to an idiolect that he had picked and chosen as well as heard, overheard and found himself echoing. It was for this reason that while he understood the sceptical notion of 'fine words', he was not inclined to say of them that they left parsnips unbuttered. Some such influences would have been at work – influences that he embraced actively – when he had mobilized the concept of the 'thingumabob' and even the 'what's its name' but had not availed himself of 'gubbins' and not (except in inverted commas) of 'Yikes' or 'diddly squat', 'jack shit' or 'nada'. There were words and phrases that remained strangers, loan words, however often he used them, as if he were assuming an alien tone of voice. 'Be that as it may' seemed like a hired dinner jacket and calling sunglasses 'shades' analogous to ill-advised Bermuda shorts on pale, scrawny legs or Dad dancing.

Some words had to wait a long time before being embraced by him. Several years intervened between his hearing and using the term 'naff'. 'Curmudgeonly' and 'codger' also had to await inner attitudinal changes before beginning to enter his speech, at least without the chaperone of distancing inverted commas. 'Blighter' never, so far as he could recall, made it. 'Hats off, gentlemen!' and 'Attaboys all round!' were mildly enjoyable archaisms. If he never quite learned to use 'I doubt it' and (later) adding 'somehow' without a certain amount of self-consciousness, it was because he did not want to instantiate the cliché of becoming in middle age like his father who 'somehow' 'doubted' many things when he did not dismiss them out of hand. RT rarely had cause to say of someone that they were an 'unsavoury' character (even less that they were 'savoury'). Indeed, another's use of such a descriptor would hint at small differences in the way the world tasted to them.

Ever the chameleon, however, he would adjust the vigour of his speech to the company in which he found himself. X might in one circumstance be 'a real pain', another 'a real pain in the neck', in another 'a real pain in the arse' and, for pub-mates 'a fucking pain in the arse'. And he learned over the years the art of the flattering introduction, the encomium to present a speaker to an audience, though he could never let go of his highlighted text. If he ever bloviated, it was because the occasion seemed to demand it and as a shy individual, anxious to appear courteous, he found occasions very demanding and was always inclined to accede to their demands. Being somewhat introspective, he never swore (which he did frequently with relish) without first ensuring that his oaths (referencing gods, genitals, or the uses to which they might be put, or a potpourri drawing on both) would not cause offence and thinking it very odd when then they did. The strangeness of the cutting-edge swear words of his childhood – 'knickers', 'bloody' – carried over into the sparingly used lewd intensifiers of his adult speech.

The world to which he had spoken was in an important sense the world that had spoken to him. His speech had never issued as any kind of first word into a prior silence. While his utterances had been 'outerances', the inner that was 'outered' was in a great part drawn from the outside world, being returned to the bruit from which it had been drawn. Discourse was a public–private partnership, where the sovereignty of the self was ceded to some extent as the price of participation in the big conversation. Even so, his, like everyone's, had been a unique voice, the site of a singular confluence of influences that had made the consciousness of RT a unique take on any one of those influences and on the common reality he had responded to, grasped, appropriated, engaged with, retreated from, by means of the movements of those now unmoving lips.

The lateness of his contribution to the history of talking meant that what he said was often self-reflexive. He could, for example: argue over the pronunciation of the word 'pronunciation', and echo the mis-'pronounciations' of others' pronunciation of this word; for the sake of fun, play with cohorts of words and buzz words, phrases and catchphrases, neologisms and archaisms; and dance with discourse – punning, rhyming, assuming accents, speaking in fragments of other languages, using a vocabulary whose words were more often counterfeited than coined. He enjoyed the embarrassed reference to certain clothes as 'unmentionables' – a pale echo of a more prudish age in which the reference to the undergarment called a 'shift' had caused a riot in the premier of *A Playboy of the Western World* on the grounds that it libelled Irish peasant girlhood – because they were thereby (un)mentioned. Quite early, he came to enjoy jokes that traded on a potential confusion between the use of words and the mention of them in inverted commas. '"Constantinople" is a very long word. How do you spell "it"?' made him laugh at his fellow eight-year-olds who began 'C-o…'. The pleasure he enjoyed (and, who knows, may even have given) through misusing linguistic registers – most often inappropriate precision, abstraction, and pedantry – contributed significantly to the *joie* of his *vivre*. Unable to resist this kind of play, he might have observed thus of his dead body: 'Thy labials are zipped' or 'Thy gob is eternally sealed.' Being forever on the look-out (or look-in) for jokes could be dangerously frivolous, dividing his attention when its undivided whole was mandatory.

He had had many interlocutors, real and imaginary, present and absent, living and dead. Chance-met strangers, friends, acquaintances, colleagues (peers and bosses, juniors and seniors), all had had their share of his talking and listening linguistic attention. Any calculations as to how many words and how many modes of speech – kinds of speech acts, with their different locutionary, illocutionary, and perlocutionary force – would be among the least reliable addition to the dubious statistics that already litter this meditation on RT's life. Suffice to say 'very many' of both and that he was more inclined to mutter than to hector and to whisper than to shout.

His most constant interlocutors were his family – his parents, for a period his siblings, his children, and above all his wife, with whom he may have exchanged – well, give it a go – over 3 billion words if they had talked non-stop day and night. But even if they had talked at the standard speed for only 1 per cent of their married life, 30,000,000 words would have passed between them. Agreements, disagreements, murmurs and mutters, and chatter and chitter, declaring love and affection, comparing impressions, categorizing, sorting, judging, fondling the multi-mediated world together, interlocking monologues, learning to disagree in a non-disagreeable way sometimes by assuming a funny accent, sharing stories, finishing each other's sentences, consonance, dissonance, performing alone, as a duet, or as parts of larger ensembles round the kitchen table, at a formal dinner, in a meeting...

His most verbose and tireless interlocutor was, of course, himself – an elusive creature in part distributed between the crowds of others in his head and in part separate (to a lesser extent than he imagined) from them. His was a monologue he could not escape and yet he tolerated its unremitting flow, far better than he put up with others' pandiculation. Yes, he told himself things that were no less empty than the pointless practical information X, pauseless as a gargoyle, would share with him, or the complete account by Y of the successful bowel action of a doted-on grandchild, or the minutiae of Z's day's work and what others thought and said about it. But the interminable logorrhoea of his own airless gassing, unlike the gas exchanges of a chance-met airhead (with no apparent deflation on either side), had been entirely acceptable – perhaps because it did not demand or interrupt his attention: it *was* his attention, or at least a mode of it.

Commenting, connecting, clarifying, guiding, reminding, justifying, daydreaming, rehearsing: these were all woven into an unremitting mouthless voice-over (or, perhaps, voice-under) that seemed, when he thought about thinking, to be located somewhere behind his eyes and in front of his neck and beneath his, first mophaired and later naked, cranium, though no one else (even if they pressed a stethoscope to his head) could hear it, notwithstanding that it sometimes assumed a tone, an accent, as he went over 'in quiet' – arguments, jokes, and anecdotes – that he intended later to say out loud.

Sometimes he was a mere passenger on a train of thought that seemed to have its own momentum. He would awaken at a station that surprised him and this occasionally prompted a backtracking: 'How did I get to be thinking about this?' And he would be momentarily disturbed at the lack of intention directing the flow of events in this, the most self-like, mode of ado. At other times, he would drive his thoughts, deliberately concentrating to a particular end, a concentration that could be disturbed by the sense that he was concentrating. The notion of concentration, of recalling himself to a centre that was the agenda he had set himself, was deeply strange. He had, as it were, to operate on himself, to maintain himself in a position, like a toddler on a wobble board, where he could be the passive recipient of the solution to a problem, or of a name he was trying to remember. How often he had referred – with insufficient astonishment – to the notion of 'racking his brains', without fully appreciating how, in virtue of being both the agent and the patient, instructing and controlling himself, the giver and the beneficiary, he was essentially pulling himself up by his hair – an especially astonishing feat in those later years when he had become bald.

Indeed, the very notion of remembering or recalling as a deliberate activity was deeply mysterious if only because it required of the one who remembered that he should have some idea of what it was he couldn't remember, of how to position himself to be ready to pounce when it rustled in the tangled undergrowth of the half-forgotten. In pursuit of his quarry – and how many of his conversations with his co-ageing friends had been punctuated by hunts for forgotten proper names! – he would mobilize connotations (the face, something the name-owner had said) to bring himself closer to the denoting term. And there were moments when he was a living self-contradiction, close to the classical circularities of the philosophy of universals where, in order to be able to understand a general term, it would be necessary to be able to identify items that would give that general term its extension. This internal pointing at an absent target, and racking one's brain to reach it, was something he had taken far too much for granted.

The ways in which he talked to himself were countless. Deducing, calculating, questioning, following a hunch, cultivating doubt or scepticism, or active uncertainty, the disciplined wondering of inquiry, of envisaging, putting himself into others' shoes or an imaginary situation, the more spontaneous ruminating over wounds and insults and the iniquity of the world, running through scenarios and their consequences, preparing for making a case – this was as ordinary to him as walking, running, eating or emptying his bladder.

And the kinds of items that it had operated with were also taken too much for granted: realizing that it is Friday, not Saturday; acquiring (at a surprisingly early age) the cognitive structures that enabled him to use (and to feel the referent of) the concept of the 'soppy' or (much later) the category of the 'naff' (though, as already noted, he felt that the use of such a word belong to others in a mode he did not entirely approve of); being able to understand the tone of something overheard on a midsummer evening in the Cotswolds: 'I rather assume that…' Such were the groins and coigns in the great cognitive architecture of his life.

The life within was on the border between order and chaos. The succession of mental fragments happening inside (in a very problematic sense of 'inside') that lifeless head had seemed like the drops wriggling down a windowpane. Uninvited elements were constantly gatecrashing his monologue, images and earworms that simply occurred without being thought into being. And yet he had been able to concentrate, to marshal his ideas, to complete an email, a letter, an article, a book, as if thinking was as active and controlled as walking to work.

If, on looking back at his life, he tended to overlook the (at a guess) 5 billion or more words he thought to himself, this was probably because they were background, because the flow was incessant, and only a small part of it was transformed into sculptured air. Thoughts did not seem to count because they were ghost actions and they did not have a place in the outside world. This was not entirely an accident. Although he had met people who seemed incapable of an unspoken thought, even these dreadful pandiculators (to be avoided at all costs), who treated contributions from others as interjections that should be brushed aside as interruptions, disclosed only a minute percentage of their endless, unbroken soliloquy.

So much for the mouthless mouthing, the lipless muttering, that had accompanied his hours as faithfully as did his shadow the body now mute and thoughtless in bed.

The speech issuing from his mouth was his entry ticket to the mighty discursive communities of his various worlds. These were greatly enlarged, though perhaps not enhanced, by technologies that propelled words across space and time. Shouting with cupped hands, toy phones made out of cocoa tins and string, phones linked up by wires, and mobile satellite phones, had appeared at different times in his life. Very few of his hours had been free of voices from afar courtesy of the tannoy, the radio, the television and the cinema. For much of his adult life he had been hounded by unsolicited music that made thought and conversation more difficult. He had been a beneficiary above all of the fact that the human spirit could be exteriorized without breath, its expression untethered to any mouth: of writing.

The catchment area of his thoughts – and of course of that of the species to which he belonged – was hugely extended by the paper, pen, and ink that took down words and sent them to places where they would not otherwise have been heard. Quill pens permitted ideas to fly beyond the spatial and temporal reach of the voice of the individual in whom they had originated. Writing set down the temporal unfolding of thought (of report, of description, of narrative, of argument) into paginated space, holding together the succession of ideas, of shaped breaths, in a succession of written lines, chapters, volumes, that were nonetheless coexistent so that they could be returned to again and again, with beginnings and ends (of sentences, paragraphs, chapters, books) visited and revisited as required. It was a delicious paradox that soot, only slightly more aristocratic than dust, could lend its blackness to the enlightenment of thought brought into plain sight. That it could inscribe colourful adjectives on pages, equivalents of those exotic birds that had flitted between the leaves of the forests from which the paper had been derived.

His early years had had much to do with ink – and with blots (two-dimensional tarns that should have been unravelled into characters) on paper and on inky fingers. The pen was a kind of proxy, if stilted, voice, secreting graphemes as it danced across the page. Sometimes the opposite poles of RT – the organism and the thinker – collided rather directly, as when sweat dripping from his forehead blotted a written word. The journey away from the organism continued as the pen was replaced by typing – with letters tapped into existence – and by word processing, where the words were out of reach of anything that would spot or blot. The journey was completed with sentences committed to the virtual space of The Cloud. Here his thought had a potential immortality disconnected from this mouth withering in silence.

His farewell to language may have been sudden – an abrupt onset of aphasia or loss of life. Or it may have been gradual: the last lecture, the last sustained discussion, the last kind or testy comment, the last time he used the word 'God' or toyed with the idea of any kind of deity, the last time he said 'I', the last time he used the definite article… We of course place too much emphasis on the last words, as if there were a special authority in the dying utterances of someone who is sinking peacefully or in agony. Goethe's 'More light!' seemed satisfyingly appropriate as did the dim-witted George V's 'Bugger Bognor'. But the poet's final utterance was more likely to be 'I think I am going to be sick again' or (an atheist's empty) 'Oh God' than something that articulates a luminous, summarizing glance encompassing his life, the self that lived it, and the world that he lived in.

At any rate, he has said farewell to RT-the-Word. He has told his last truth and his last lie, made his last helpful or unhelpful suggestion, cracked his last joke, and argued his last case. He has felt for the last time how the space into which his words were uttered might itself become audible, courtesy of the hum of a fridge, an open window and a bellying curtain pregnant with the evening air, or the sound of a distant dog barking, so that he could sense the great silence placing gigantic inverted commas around the sum total of his speech.

There remain only his two lips, pressed together but unaware of each other, and nothing more to say. They have kissed themselves farewell.

twelve

Inter-space: Together & Apart

A Thing Apart

No man is an island but every corpse is.

Behold the late RT, disempowered, dispossessed, and voiceless; alone beyond loneliness, solitary beyond solitude, as separated, without any sense of separation as a pebble. The weight of the body and the resistance of the bed are no more in dialogue than the mutual pressure between the bed and the floor on which it stands. The objects are connected only in the eyes of the living, who neighbour them. The mottled skin, which no longer delivers intelligence about the world, defines a boundary that is no one's boundary.

His 27,000 yesterdays had seen him connected, interconnected, and disconnected. His consciousness had reached out to a world of persons, places, and things, and they had reached into him. He had been a site of meanings whose locus was neither inside nor outside him: RT was a between-being. Now, as the saying goes, he is 'no longer with us'. He is no longer with himself. He is no longer 'with' at all.

From this utter privation, we may recall, or imagine, the extent to which he was linked with, or divided from, his fellow worldlings and, indeed, the world itself, to what degree he was dissolved, or enisled, in the wider sea in which he had lived his life; consider the 'us' he is not longer with and reflect on all the varieties of 'with'; to remember RT as confidant and stranger, lover and fellow passenger, insider and outsider, at the centre and on the edge, participant and onlooker, bystander and protagonist, on the bench and in the ruck, leader and straggler, in charge, in another's charge.

I and Thou (1)

His life, of course, began with union; a unique link, never to be equalled. It is commemorated in the untidy, sunken little plaque still visible on his abdomen, and relatively unaltered with time, though the terms he used to refer to it changed, as he grew more sophisticated: tummy button, belly button, navel, and *omphalos*.

A little navel gazing is called for.

It was as part of the body of one member of it that he joined the human race; as an outgrowth of the first human company he would ever experience. Confected from items that had had little to do with one another, RT had gathered into a distinct, self-directed being, a being who was a matter of unceasing concern to himself. This primordial union with another, forged when he was a clump of cells multiplying in a uterine darkness, preceded his existence as a conscious being. The world where the woman destined to give birth to him stood up, walked, and sat down, talked with his father, felt him kicking and worried when his legs were still, did not reach him even as a rumour during his nine months of auto-fabrication.

At full term, he had been expelled from this Utopia where his needs had been met before they were experienced. The umbilical cord, roping together the two organisms, was severed by the midwife's scissors in the first minute of his life. And so he was delivered to the world as a distinct entity, sealed off by his all-enveloping skin and by all-engrossing needs that he himself experienced directly and others only inferred. Without this primordial severance, there could not have been any experience of 'being together', of 'with us'.

After the scissors had done their work – leaving that nubbin of flesh as the permanent reminder of the contingency of his origin, and the consequent necessity of his mortality – he was aware of his mother as a presence. His was a world in which he could feel her warmth; bask in, and then respond to her gaze with an answering gaze interpenetrating with hers; smile at her smiles; and eventually respond to her voice with a voice of his own. His distinct identity was thus dependent on links. Just as a shared gaze requires two separate viewpoints, seeing different things, so apartness and togetherness – apparently distinct or even opposite – are themselves inseparable.

It took time for this first 'I' and 'thou' to crystallize out of the whirl of his being in the world. At first, she was not someone in herself, a face facing his. She was much more: a condition of his being, the landscape and weather of his existence, a warm mountain running with manna and comfort, an absolute necessity, whose withdrawal was Hell and whose presence was the possibility of bliss. She was more than someone and less than someone, until eventually she crystallized out of the neonatal delirium of dapplings of light and shade, of sound and silence, of satiety and hunger as Someone in Particular, who loves, judges, encourages and obstructs you. Exchanged glances, smiles, games, and words braided new bonds and at the same time opened up new distances: he and she differentiated into 'me here' and 'you there'.

One day, he stood on his own two feet. He started to walk – away and back, to and fro, ever further away, ever more slowly back, away more frequently and absent for longer. His physical presence was increasingly represented by a proxy presence in words. The apron strings were cut by instruments more complex than the midwife's scissors – skills of independent living, modes of being 'away'.

He developed his own distinct viewpoint, something that had begun literally with his gaze on the world and glances shared and not shared with his mother. The child, boy, youth excited by and fearful of possible worlds ahead, and rejoicing in a growing freedom, was doubly separated from the parent who dreaded the separation, who said 'Take care' and worried that the world in which he was finding his own path would not take care of him.

The one who has set out and the other who is left live largely separate lives, the one in the rain of loss, the other in the sunlight of discovery, crossed by intermittent phone calls, dutiful letters. Growing up would, it seemed, always be growing away. Away, too, from his siblings whose navel-strings had shared the same hitching post, who had been factored into being in the same cave of making, and whose childhoods closely co-habited with his own.

Yes, they would from time to time be ambushed by emotions belonging to the old closeness, by forgotten sympathies and antipathies: each remained to that extent inside the other. But there was a relentless springing apart. He put out from the shore, swam beyond their shared depth, and looked back from greater distances to a shoreline becoming ever more remote, his longing to be free counteracted by his wish to be loved, to be embraced and yet to walk unfettered. The gap between farewell and hello widened and that between hello and farewell narrowed as he was launched into the wideness of the human world where togetherness was always dappled with apartness and apartness with togetherness, and separations that were as poignant when barriers were closed gently as when they were slammed shut. One day he would marry and she would hear her child's own children – the product of a life she now knew only through long glimpses interrupted by longer absences – calling her 'grandma'. Other distances – his roles, changing relationships to relations, preoccupations, offices, the carapace of importance, duties, and seriousness – opened up. And then she would die and the world would lose the place where he had begun, a powerful sign of his passage from necessity to contingency, the asymmetry between the Special Where of his start and the Anywhere of his end.

So much for the essential, necessary, and irreplaceable relationship of his life; the one in virtue of which he had existed. What of the others?

We, He, She, You, They

RT had rarely been alone in his 300,000 or more waking hours. It was as a member of a variety of assemblies, one face among several or many, that he had passed most of his life. He milled with milling crowds; queued with strangers, united or aggregated by some shared need, adding one unit to a pure series defined by abstract categories ('needing vaccination', 'requiring information', 'waiting to be served'); elected to be an atom of an audience sharing a common interest; was one passenger in a train packed with several hundreds, occupied one seat in a fully booked plane, or stood in a bus with standing room only. During his childhood, boyhood, and youth, he spent much of his time as one head in a row of heads, itself set between rows of heads, arranged with geometrical precision like crops in a field, one of the more attentive members of classrooms and later lecture halls packed with grudgingly or intermittently inattentive youths. He had at various times joined ensembles ranging from teams (closely knit, or ad hoc) that played or worked or volunteered together, guests at a dinner party, or a gala dinner, to crowds at receptions, to duets, trios, and quartets of friends, gathered for a gossip, or a laugh or to plan some future event, or to share confidences, sorrows, fears, and joys. He had marched in ranks and processed in processions. He had not infrequently got together with others for no other reason than to participate in a get-together. Or for something almost as non-specific such as bonding, a chat, a catch-up, or a laugh.

The modes of togetherness varied in their richness, depth and intensity. The class of ''66 Reunited', a group of fellow hobbyists or campaigners, the members of the department, colleagues, juniors, and seniors, assistants and bosses, mates and enemies in the platoon, were united by the idea, informal or underpinned by contracts, casual or dutiful, of something they had in common. He was a joiner and a non-joiner, one who participated and one who did not show up, usually the latter, preferring solitude. He believed he was discerning, when he added his one to others united by a common purpose, his name to a list of signatories, and avoided swelling a herd gripped by deluded beliefs, false consciousness, or evil intent. Being together with others sometimes awoke a surplus of solidarity, a feeling of collegiality, expressed in banter, joshing, complaining about bosses or juniors or shared burdens, the emission and receipt of *bon mots*, and a variety of ways of celebrating *Mitsein* for its own sake, quite apart from any convergent goal it might serve.

Sometimes togethering was spiked with competition, envy, irritation, even spite. More often there was benign indifference, a distance that came from the mere fact that the meetings were, for each of those who met, located in a different history, situated in life-tracks that intersected but did not intermingle or interweave. He and A. N. Other, washing their hands in adjacent sinks in a public toilet, were coming from and returning to different events, situations, worlds. The temporary links affirmed by greetings and broken off by fare-wells, bracketed by handshakes of meeting and of parting, by Hail and Farewell, lost their significance when the reason for the meeting was achieved and the rationale of partnership expired.

He had often shared in the transient unity of audiences in audito-ria, occupying Side Circle Right B3 in a hall that could accommodate several thousands who had come for the same event. Even though there may have been widely different reasons for attendance, the disposition of still bodies and attentive faces pointing in the same direction gave the impression of oneness. This was soon exposed as superficial and transient – as when attendees looked away from the shared object of attention, the performance, and exchanged mean-ingful glances, or, enraptured, turned to each other's applause in search of approval for their approval, or resented the queuing for the cloakroom. In smaller groups, there was (and it had sometimes been RT himself) someone surreptitiously consulting a watch or tapping text into a phone or migrating inwardly to unshared places.

The coming together and coming apart of human clusters in a world full of strangers had been something of a preoccupation for RT who, for much of the early part of his life, was, or thought he was, socially awkward. He had an abiding fear of being unwelcome. He had neither gatecrashed nor wormed into places, even when they seemed more brightly lit than his own rather ordinary parish. When he joined a group, he had felt disadvantaged in virtue of being out-numbered and judged by more stares than he could return, by the gazes of people more confident than he, who seemed more at home in any given *quartier* in which he found himself. There was the particular difficulty, already noted, of taking his leave – seizing the correct moment, remembering his lines, finding the path between laughable formality and an eyebrow-raising casualness. He had especial problems with seniors, even when he was slightly senior himself, the awkwardness of his relations with the primordial authority figures and judges being replicated in adult life.

Or at least until, in his thirties, he suddenly found himself fluent and (comparatively) unjudged. Then he was able to talk with anyone – fellow passengers round a table in a train, eminent members of his profession, the dying *in extremis,* members of an audience he had lectured, the famous and the obscure – giving them the impression they were being listened to, tuned into, that they were immersed in a warm bath of personal attention, and giving himself the impression that he was engaging with them. As for parting – to return to his work, or to leave the train, room, or building – this was no longer a wall to be climbed but a chance to exhibit even more fluency and courtesy.

There was an obvious reason for this acquired at ease: with the passing decades he brought more CV, as well as more practice, to the moment. The current account of the exchange was smaller in proportion to the growing deposit account of his life hitherto and what, on paper, and hence in his sense of himself, it amounted to. The discomfort of being with others – even the sense of deserts of time stretching ahead at the beginning of a long meeting, and the joy of being off duty when the meeting was over, when he would be free to stop being polite and could think to himself – became less acute.

All of this palliated the anxiety that had preoccupied him when he was younger: that he did not know whom or even what he was talking to; that he was as ignorant of his interlocutors as they were of him; and that their presence (so long as nothing embarrassing, or annoying, or substantial took place) had passed through him like water through sand. In proportion as he was socially at ease, so the changing circumstances in which he found himself were less stingingly present. The succession of formations to which he contributed was swept up into the dance of the hours. Just as the properties of a tool, its opaque materiality, did not obtrude on him so long as it was in working order, so the little challenges of being-with were dissolved by his fluency. Those with whom he spoke were sufficiently intelligible for the situations in which he met them. Cooperating with a colleague, sitting side by side with a fellow passenger with whom he exchanged the occasional remark, sharing impressions with another guest at a wedding – under all of these circumstances, the strangeness of the stranger was counteracted by the legibility of the shared situation. Perfect strangers were imperfectly strange because his knowingness patinated with over-familiarity what lay outside him; it muffled singularity, and extinguished bright difference. The unknown and unknowable hinterland did not obtrude, or it was lost in the plainness of plain light.

Even where there was no 'job to be got on with' he was still able to tolerate the absence of true togetherness when he was with others. The shallows of the interpersonal, however, were more easily tolerated when there was competition. Others were reduced to bearers of potential scores that were to be outscored. Or anger, which took things personally, and was satisfied with knowing very little of the other, except as the culpable source of what had made him angry. Indeed, the less anger knew of the other, the better because the less there was to stuff into any verbal receptacle such as 'clumsy lout', 'selfish bastard', 'cheeky bugger', 'flaming idiot' (why flaming?) to contain the object or subject of a self-justifying rage. Rage is satisfied with the Other as the referent of an abusive epithet. In conflict, he was uncomfortably aware, both parties could rest assured that the other was the number one agenda item in the meeting: to that extent there was a meeting of minds, even though the minds were looking at the common situation from irreconcilable viewpoints. It was possible to be more continuously and intensely together through mutual irritation and resentment (often justly described as 'burning') than through more creditable modes of awareness of others.

But anger would usually pass and he would look back on its compelling importance with dismay and surprise at himself. The person who had stepped on his foot, the idle sod who had dodged his share of work, the opponent in the long-drawn-out dispute over something or other, would all be forgotten. He saw that the tendrils joining him to the other members of the endlessly changing ensembles in which he found himself were of the most fragile kind. The cliché was largely true: each in the crowd enisled and alone.

With certain important exceptions.

I and Thou (2)

So much for 'they' and 'he' and 'she' and the various modes and clusters of 'we'. But there had been, RT had believed when his body had had beliefs, something much closer, in which togetherness triumphed over apartness: two minds with but a single thought, two hearts that beat as one; his own and that of the now stricken survivor, the sad heart and sorrowful mind that circles round his mindless head, his stopped body.

When he had met with friends, he and they had seemed to be together rather than apart mainly because they were literally closer than they had been when they had planned the meeting over the phone; they had got together in a get-together. A arranges to meet B when they are separate, or are too busy to address each other. Compared with this, being at the same table, enjoying time specifically set aside for each other, is an obvious narrowing of apartness. This belief is reinforced when confidences – 'Don't tell a soul! This is strictly *entre nous*' – create a shared interior in a world of outsides. Was it possible, however, that this had been only a distraction from a more fundamental, more dismal truth, revealed when he had looked into the face of interlocutors and remembered that he had no idea what they were really thinking or feeling? Or that, even if his guesses were roughly right, he had no way of being sure of this?

There was a more profound darkness at the heart of togetherness: he had no idea what his own presence was like. At the very basic level, what did he look like when he came into a room? There was an incurable invisibility to himself in his visibility to others. How did his facial expressions – visible to everyone in the room except him – strike others? Did it not happen from time to time that someone might say to him, 'What is that look on your face for?' when he was not aware of having any look? What of his bulk, or lack of it, his clothes, his general demeanour? Would that silly grin look sillier or less silly in the eyes of others? Perhaps someone over there was classifying him as an unsavoury character. He was the one person he could not see in the room, the carriage, the pub. His looking face had no idea how his look looked.

There was, of course, no question of a general answer as to how he looked 'in the eyes of others' because there were different eyes and different others and different situations. The idea of an observer who captured the average appearance he had in the average eye, or the just eye, had only to be spelled out to be exposed for what it was. (Hence the fakery at the heart of novels with their 'perceptive' authors.) And how did he *sound*? His assertions, his accent, the proportion of the dialogue he took up, the volume, tone and timbre of his voice, his jokes? Take the acoustic properties of his voice: did they grate on some ears, please others, and for most auditors simply act serviceably to convey information? As for his 'accent', its being an accent and the expectations that that accent conveyed was a function of the unimaginable consciousness of those who spoke his native language differently from himself. Was there a specifically Northern tinge to the consciousness of someone who had a Northern accent? Of course not. But accents tuned expectations.

There were questions about his impact on other senses that he could scarcely bear to entertain. And yet more complex questions relating to the higher-order aspects of presence, such as his mode of attentiveness, his sincerity or insincerity, his standing as a *mensch* or as an idiot. Whether and under what circumstances he had charm and of what variety – roguish (probably not), quiet, seductive (unlikely), fake, sinister (unfair) – if at all. What was it like to have him around, to be his junior or boss, to sit next to him in a train, to be two tables away from him in a restaurant? What was it like, in addition, to experience his presence through his absence, through his not being available, not turning up, keeping someone waiting? Or when he cast a shadow that made others jump, or made a comment that was not quite heard for those who struggled with his mumbling?

There was no amending this ignorance because much of what it was like to be in his presence lay below the level of recorded, recordable, reported impressions. The gaps in his knowledge could not be filled by a handful or a bucketful of adjectives. Off-the-shelf epitaphs were not customized to his singularity or its interaction with other singulars out of which presences, and the impressions of presence, were formed. The adjectives were not helpful for the additional reason that he had never been anyone other than himself to calibrate them. Everything, ultimately, was mediated through himself.

At the heart of his sense of his own presence in the world therefore was ignorance as to how others sensed his presence: to be with others was to be a blind spot in a shared visual field. That presence, try as he might to manipulate it, was ultimately in the keeping of others. If he was more than usually allergic to unjust criticism it was because it reminded him that he was impotent to control how he was judged. There was feedback, of course, but patchy, incomplete, and filtered or provoked by other considerations than the sharing of the truth about how his entry into shared space warped the social gravitational or levitational field.

This ignorance had been compounded by his inability to imagine what it was like to be those others who tasted his presence: how they tasted themselves. At the heart of the matter was the fact that when there was a meeting between A and B, the meeting for A was a meeting with B and the meeting for B was a meeting with A. Two meetings, albeit simultaneous, but nonetheless quite different. A could see that B was looking but was not sure what she was looking at and even less sure what she was seeing, even when the question related to something as uncontentious as a busy street or as simple as a tree. 'A relates to B' does not equal 'B relates to A'. This rule of noncommutativity in human relationships is ubiquitous.

None of this had concerned him for much of the time. It did not stop him getting on with people, at least to the extent of 'rubbing along' fairly companionably without either party digging deep or even feeling that he was being confined to the shallows. Common purposes swept aside the questions about the 'who' of the other. It was against this low baseline – where he had been used to settling for something a long way from true closeness, a real co-presence – that sexual love had seemed to abolish apartness. He had been so used to being separated from others by space, time, his and their comparative indifference (notwithstanding outward demeanours that were thoughtful and attentive), profound mutual ignorance, and justified lack of certainty as to how his own presence was being experienced, that certain modes of togetherness looked like true connection; that something more intimate, with desires recipro-cated, secrets shared, and a mutual concern, such that out of sight was *in* mind, had seemed to close the gulf between RT and at least one other human. It was as if the elsewhere inescapable reality that when people meet, they come to each other from different lives lived in fundamentally different bodies, forged and continued in overlapping but still separate worlds, retain different viewpoints, pass different judgements, have different expectations, different hopes, desires, sensibilities, different triggers for joy, irritation, dismay, and delight, had been suspended.

Happily these uncertainties were forgotten, or obscured, even in the foothills of the relationship, by the joy of being together for the sake of togethering: hand-holding, kissing, cuddling, saying a joyful 'Hello' in the middle of a conversation. By the delight of sharing, comparing, contrasting, agreeing and disagreeing on impressions. 'Look at that light over there!' he says, raising her hand clasped in his as he does so. But the fundamental limits on togethering were not suspended even in these very special circumstances. He and she are familiar and unfamiliar with different things, they have had and are having different struggles, and the dappled patterns of 'home' and 'away' are differently distributed over the world they share. Indeed, in order that there should be a *between* them, they have to be irreducibly two. Sharing is consequently episodic and even in those episodes, oneness is an unachievable asymptote. There can be no fusing of minds (after all, 'to mind' is to be both distant from and in contact with), no merging to a single viewpoint, preoccupation, mode of continuity. Even if their two minds did have a single thought, that thought would be located in a different network of other thoughts, experiences, and preoccupations. Oneness without barriers was actually at odds with the spirit of romantic love, which required the distance, the mystery, the otherness (to use the jargon) of the other.

At the very basic level, when they are side by side, he was to the left of her and he was to the right of him. He was always over here and she over there – and *vice versa*. It is a question of two lives, after all: she is in the garden, he on the train; she on the train aware of other passengers, he awaiting her at the station, getting cross with the idiotic opinions on the radio.

They enter into and exit each other's lives, while there are no entrances and exits so far as the self is concerned: RT's exit from X had been cancelled by his entrance into Y. Notwithstanding his feeling at times that he was going 'out', or 'outside', away from the centre, the bright lights, into the wilderness, away even from significance, this was still an entry to another place. Even when RT said a temporary goodbye to himself as he slept, his exit from the world was an entry into another world; and waking out of dreams was an entrance as well as an exit.

So, through all the years of loving companionship, of a shared life, shared duties, shared spaces, shared experiences, shared sorrows and (much more numerous) shared joys, he had still been only episodically present to her while continuously present to himself and she had been only episodically present to him while continuously present to herself. Her toothache is a series of reports for him; or his ignorance of it is only slightly alleviated when he learns that she is in pain again. This was the law of even the closest relationship: 'Thou art intermittent and I am continuous' – intermittent even when the gaps are full of images and thoughts of thou. Or not entirely full because we are, after all, distributed between many others. After the anguished farewell, there is the person who annoys us in the carriage, the phone calls to be made to colleagues, the document, client, patient, or customer who reasonably expect our undivided attention, the book we are glad to read. We have little idea of how we figure in our nearest and dearest's thoughts, speech, and involuntary images but there is no doubt that we imagine our presence as more coherent and continuous than it could possibly be in fact.

The jealousy and suspicion that flickered round the edges of the most confident, trusting and passionate love highlighted the reality of yawning gaps in knowledge, and in mutual presence, and revealed the other as an embodied subject hidden within an opaque world whose opacity began with his or her beloved body.

Lovemaking, even at its most intimate, communicative, and joyful, did not obliterate apartness. The very term 'sex', with its generality and its universally recognized staging posts on the way to the universal climax, had always been susceptible to become a succession of transactions, and even the much-deplored commodification. If not bought and sold, it could be given in response to needs of the recipient that the giver recognized but did not fully understand and perhaps could not even imagine. Which was why it was so often traduced. Sex as a tradable commodity – an exchange of bodily fluids without an exchange of vows – the one-night stand, abusive sex, sex as violence, as possession, as an assertion of power, as conquest, as one of the spoils of military or other triumph, as the imposition of a disgusting intimacy on another who does not wish it, as something one steals from another, sex that aims to degrade and humiliate, sex preceded and succeeded by indifference, all bore testimony to the deep ambivalence at its heart. This ambivalence was unresolved even in the extreme closeness of two people bound together in a loving, delighted, respectful embrace, one body inserted into another, a closeness that did not entirely cancel the distances present in the ordinary relations between individuals who simply get along with, or shake down with, one another.

Even in the midst of mutual delight, each is separate from the other if only by virtue of experiencing different sensation, different phases of pleasures, embedded in different stories. No orgasm can entirely possess the body of the one who experiences it, even less reveal the person-haunted or world-tinted carnality of the lover who has made the orgasm possible. No desire can ever reach or grasp its object because its object is a subject who cannot be reduced to any transient state, however ecstatic, privileged or beautiful – unlike appetite which can be satisfied, because it is cognate with that which extinguishes it. The mystery one finds in the body of another, the fascination of a fold of flesh, of a colouring of the skin, is not evident to the one whose flesh it is. She does not share the fantasies of an exotic Other World to which, for him, it belongs. Unreachability lies at the very heart of desire, which desires the irreducibly different, to know the unknown while keeping its mystery intact. At best the summits of ecstasy can subside into the uplands of intimacy, affection and gratitude before each rejoins everyday life.

The very closeness and specialness of the sexual encounter may mean that the ordinary worlds each come from and inhabit are marginalized. Which is why of course sex can be bought. The trade in sex (and the free-floating lust that underpins it), the tariff for the bill of fare when bodies interact rather than worlds, a physical intimacy dissociated from any other, betrays something impersonal at the heart of carnality, where the other offers an adventure that does not have to be shared because it is remote from daily life as are the physical interactions, the warmths and coolths, pressures, textures, scents and secretions, that are associated with it. Even when each faces the other at the supreme moment, neither sees what the other sees in his or her face. The gift of pleasure, of 'Yes!', of intimacy, even when free and motivated by no other wish than closeness, leaves an irreducible gap between giving and receiving. At a very literal level, the joy of the one who is ensleeved and the joy of the one who is ensleeving do not map precisely on to one another: they are not congruent, however consensual, truly intimate, the lovemaking. There is no pure mirroring: always *mutatis mutandis*, never *ceteris paribus*. And the self at its most pleasured is at its most self-centred. While talking and caresses are together or between, perhaps, the intense pleasure of orgasm has something solitary at its heart.

The limitation to the closeness of lovemaking betrays what is true of all togethering; that he cannot know what it is like to be with him. Even in those closest moments he cannot be entirely certain what the other is feeling. He has no idea what it is like to be made love to by him – her response may be calculated out of love and generosity. It is not impossible that his moment of supreme pleasure is her moment of moderate discomfort. More specifically, what is it like for her to feel his weight on herself, to await his next move, to observe his pleasure, to signal her own? He had known what it was like to encircle her with his arms, and what her encircled body, at least the patches he was in contact with, felt like, but not what it felt like to be encircled by him. After all, there could be a profound dissociation between the pleasure he took in her body and the pleasure that she took in it. She might be as desirable, irrespective how she was feeling in herself: whether she was ill or well, randy or simply tired. It is possible to look sexy and not feel it or to feel charged with desire and be undesirable.

As he embraced and caressed, he knew that there was an ineradicable uncertainty in defining the object of touch: skin, a body, a person, a person becoming or coming closer to the part of the body that was touched. If touching had been sufficient to guarantee a deep connection between persons, then a tube train packed with the bodies of irritated commuters and opportunistic gropers would be the theatre of radiant co-presence and happy mutual awareness.

To fall asleep in another's arms is perhaps a more complete image of togetherness. But sleep is a profound interruption to togetherness. The closeness marked by shared warmth like an exchange of whole-body smiles, the clingfilm of perspiration that they joked had increased the coefficient of friction that caresses had to overcome, to which one provides the *recto* and the other *verso*, was deceptive, not only because his *recto* is her *verso* and her *recto* his *verso* but also because two can lie together who mean nothing to each other and because the film remains even as each falls away into the separate ultimate privacies of dreams or dreamless sleep. And this is all too evident to the one who remains awake, thinking about the things he usually thinks about, that have nothing to do with this bed, a consciousness with its most elaborated intentionality directed elsewhere, listening in underlined solitude to her breathing, or to her small cries as she acts out some drama in the theatre of her own mind in which she is the involuntary protagonist. He could awake and comfort her, asking her as his mother had asked of him when he was a child, 'Did you have a horrid dream?' The comfort born of physical closeness and the blessing of total attention, a sunlight dispersing the shadows, might make her feel (without explicitly believing it) that all will be well. That reassurance has to ignore a distance so unreduced that it may be necessary for one to ask of the other, 'Are you awake?' Or for them to share and compare, in the morning, the night's dreams where they had been in different places living different lives pursued by different fears.

According to Heraclitus, only the waking share a common cosmos; each sleeps alone. So let us return to the waking hours in which their lives, feelings, impressions, and preoccupations are (seemingly) shared. There are entrances and exits. She closes the door and he loses sight of her, of her immediate concerns, what she is doing and thinking about, and he returns to whatever he has to do. Worries, in particular, separate them: he shares hers, of course, but cannot live them as she lives them, heartbeat by anxious heartbeat. It is she who has to act on them, take responsibility, and be held responsible. Something analogous is true of the fundamental troubles of illnesses and pain. While his fever is visited at intervals by her loving concern, he lives it moment by moment in a transformed world. Indeed, it is because he cannot feel it that he inquires after her pain. Even in the case of a humble cold, there is the one who hears the sniffing, blowing, and sneezing and the one who sniffs, blows, sneezes and suffers their underlying cause. The sneezes that give him relief may, if sufficiently numerous, start to irritate her who has no choice but to listen to them, or await the next one like the ninth slam of a door. And there are many situations of extreme importance to each – a job interview, a surgical operation – where, however much they would wish it, neither can stand in the other's place nor even imagine what it would be like to do so.

Their togetherness might be evident on a walk when one says something, quite unconnected with their present situation, that the other was just about to say. They are astonished and delighted at this proof of their closeness: two minds with but a single thought implying two hearts that beat as one. And yet this exceptional event is proof of the more general rule that, even when they are side by side for no other purpose than being side by side on a walk that has no purpose other than a pleasure enhanced by being shared, their thoughts run along different lines and each has little idea what the other is thinking. The points of convergence reveal the lines of divergence. They affirm their togetherness by pointing things out to each other, creating bridges through the joining of attention. But the pointing out highlights the gap even when the invitation to share may be resisted. He is brooding over some problem at work; she would rather just look at what is before them and not be dug in the ribs, howsoever gently, and invited to do what she is doing already.

A more potent affirmation of unity was joint contemplation of past experience, that they alone could know. The more seemingly trivial the memory, and the more minute the remembered detail, the stronger the proof of their privileged access to each other.

While they can recall, or share, of this remembered past only the factual outline and the fact that they were together, this increases its exchange value; for the event is presented as pure togetherness, with the distance of time compressing any separation between them, eliding any source of distance. To say, in the darkness of a shared bed, 'That walk on Pentire Head 1987' seems temporarily to turn the decades of separate but parallel or braided tracks linked by cross-bridges – held hands, glances and smiles, acts of kindness and of love, phone calls, car journeys, connecting flights – into a truly shared life. The inescapable physical distances – separate continents, countries, cities, streets, buildings, rooms, sides of the bed, heartbeats – are momentarily collapsed.

But there were distances that love could not entirely eliminate, even when there were no disagreements, misunderstandings, quarrels, suspicions, or sudden revelations. And the seemingly most potent manifestation of the distance arising from the fact that they had two lives not one – anger – could unite as well as divide. Anger, that is, whose object was each other, not some third party or external issue in the face of which they felt solidarity. Hatred could be close – and certainly closer than distracted cohabitation; mutual irritation or burning resentment could create a more intense and sustained togethering than contentedly rubbing along or even deep affection. To be angry with someone one loves is to suffer a malign inverted intimacy. The fury fans the flames of self-awareness as an awareness of the other. The violent argument, with accusation and counter-accusation, expressing hurt, indignation, and disappointment, may come to seem like two flames joined in a single fire; at any rate, it seals off the couple in a place in which they have eyes, armed with daggers, only for each other. And there is the wonderful release of reconciliation that makes a happy miracle of ordinary togetherness.

But each knows that there are fundamental limits to any relationship from the fact that *relata* have to be in many sense external to one another. Each has to bring a biography, a self, to the party, something that the other loves because it is different, because it is only surmised. There is an irreducible she–I and he–I in the irreducible thou–I, and even an irreducible It in the irreducible She and He, as there was an irreducible It at the heart of the livingness of the I.

That in the Other that lies beyond the reach of the Self is inadequately captured in the idea that there is a 'something' that 'it is like to be you'. The something is a tone of self-presence, an awareness of the world, and a mode of moment-to-moment coherence. Even if he had wanted to have told her what it was like to be him at any moment, he could not have done so because there was no such thing that could be summarized, no account that could bring together a train of thought, a hint of indigestion, satisfaction at making a point at a meeting, being distracted by a telephone conversation, worrying about a deadline about to be breached or a patient going down the pan, and looking forward to a holiday on which their togetherness would be less interrupted.

He knew, nonetheless, that the romantic pessimism of thinking that each was sealed up in a world-for-one was nonsense. His very actions and their intelligibility to himself drew their sense from a shared space. If RT had been an island, he had been dissolved in many invisible oceans that made up the communities to which he belonged and in which the tissues of his self had been forged. Apartness was inseparable from deep togetherness. Those others, whose other he was, and he himself, were soluble fish in the ocean of human life.

Which was why each was utterly familiar to the other and there were times when he felt that he was known too well or, as he might have put it, 'only too well'.

No man is an island.

Nature

'No man is an island but every corpse is.'

Let us see.

At the heart of RT's very existence there was something close to a contradiction. He had been part of nature – it was after all natural processes that had brought him into being and natural processes that had been his undoing – and yet he was sufficiently apart from the natural world to seem to be acting on it as if from the outside, and living his life in a human realm that was in many respects opposed to the natural one. There were other ways of capturing this antinomy; for example, that in order to shape his life according to his wishes, a vision or an ambition unknown to nature, he had had to exploit nature's laws. He was, he felt, the author of his actions but for them to happen and to have their desired consequences they nevertheless relied on a boundless, seamless causal net. Without the laws of nature operating in every part of his body and governing without exception the unfolding of events in the theatre of his life, he would not have been able to live his life in accordance with his ends and aims. He needed to be inseparable from a coherent, seemingly causally closed, physical world so that he could intervene in the world as from the outside.

The coherence of that world was astonishing. Superficially, it was a matter of one thing after another, one thing next to another. But there were habits and connections visible to the most casual gaze.

Thunder followed lightning, drops of water set circles moving over the surface of the water to which they had donated themselves, what went up came down. The material world remembered to behave in a connected way. The breeze that riffled the pages shook the tassel that marked its place. The rain shower on hot pavements unpacked itself in a lovely mist, a stone dropped in a pool made reeds tremble and forced the water boatmen to struggle a little harder to achieve their dimly conceived goals, a shout remembered to echo in the right tone of voice, a bench in a hot little railway station cast a shadow divided into slats by brilliant sunlight, an explosion made the dust jump on a window sill miles away, and the wind orchestrated the dance of the trees and flowers.

Moreover, as his fellow humans had looked deeper into the world in which they found themselves, a different kind of coherence emerged. Laws were discovered and they in turn were found to be manifestations of ever more general laws. The habits of matter that made it possible to make and keep appointments, for him to will his various modes of togetherness, to have, to share, and to communicate experiences, were astonishingly consistent and universal. Beneath the casual, even patchy, surface order, things were co-ordered, coordinated, gathered up in a dance of co-variance. Thus the unwilled coherence that permitted him to impose whatever coherence he could on his own life and to conform to, and rebel against, those mores in accordance with which he regulated the lives he lived in different social circumstances. If Newton's Second Law had not been upheld, indeed unbreakable, if f had not equalled ma, his private, public, and physiological life would have been chaos; more precisely, it would not have been.

So, throughout his life, for all that he felt he shaped it in accordance with his own idea as to how it should go, he relied on circumstances he could not have created. The most fundamental and universal of those circumstances was his functioning body, aligning its law-abiding processes to selected aspects of a law-governed universe. Multitudinous physiological mechanisms, of which he had little knowledge and over which he exerted even less control, were necessary for his intentions to be realized and agency (primarily but not exclusively his own) to permeate the story of his life. *Making* things happen in a world in which things 'just happened' was a brief power granted to items like himself that mattered to themselves and wanted to make their little parish of the universe dance to their own tune, and this power was rooted in happenings largely hidden from him.

The coherence of RT's now stilled body had hitherto been many-levelled. What happened in the individual cells was *orchestrated*, adding up to particular functions, including the functions of keeping alive, of remaining importantly unchanged in the face of necessary change. This coherence was even more explicitly evident in the cooperation of the billions of cells in the organs responsible for keeping his blood circulating, his fluid and ion content balanced, his tissues oxygenated, and his body upright, mobile, able to act to ensure its own physical and even social survival. The union of this vast assembly of bodily processes had made it possible for him to be an upright, thinking self, to be a more-or-less unified someone aware of being upright, and of thinking.

The great discovery of the century in which he had been born was that, at a fundamental level, the world was more continuous, less differentiated, than was apparent to him, or to anyone else. Separation in space and separation in time were only space-like and time-like separations. The order in which events took place was not intrinsic but dependent on the relative velocities of the frames of reference from which observers observed them. There was an over-arching viewpoint in which all events were coexistent and Becoming was frozen. The casual unfolding of the world, seen in the soft causal connections between things, in the coherence of an afternoon in which a breeze that makes a hedge sway and windows rattle and papers tremble, and water acquire goose-flesh, was the mere surface appearance of a deeper order. The eyes that had drilled beneath these appearances had reached a level of understanding according to which the fundamental components of the universe had no clear boundaries or locations. Under the penetrating gaze of fundamental physics, discrete objects and the fields of energy and forces around them melted into a universe that had a single wave function gathering up all properties and relations. It would seem as if everything was one and connectedness was the ultimate truth beneath surface separation.

No corpse, it seems, was an island and nor was any island.

If this was a beautiful idea rather than a mere mathematical fancy (and he had believed it was the latter rather than the former) it was not one that afforded any of the consolations that might be expected to flow from it. For without apartness there was no together. There would be no 'I' to live out his life and no 'thou' to make it meaningful or 'they' to make it liveable. The universe as experienced by RT, the theatre of his life, was seen from the point of view of a distinct individual acting upon it as if from without. This discrete entity was located in a particular body that divided the world into parts that were near and far, useful and useless, promising or threatening, precious or without value. An eruption of self-consciousness into a seamless material world, RT had been the centre of a parish, or a succession of interlocking parishes, interacting with other centres of near and distant parishes, populated by discrete items and discrete people. As an agent with wishes, needs, hopes, ambitions, virtues and vices, he had been a continuous self, interacting with intermittent others, subject to the dappled togetherness and apartness that had made his life both tragic and bearable. Of this physics had nothing – least of all anything reassuring – to say.

And this is where the paradox returns. The emergence of the viewpoint, of the embodied subject RT, had itself been dependent on the material properties of a body subject to the laws of nature. The universe that had forged him, granting him a biography in which he had had something of a hand, permitting him to participate in a dance of togetherness with those who now mourn him, did not seem to have room for a discrete, self-aware entity such as RT which would shatter its continuity in accordance with his own interests. For that, if no other reason, it had been inevitable that it would also break him in its turn and digest and dissolve him into the mindless unity of Being.

No island is an island but humans are; temporary islands, as their mourned corpses testify.

And & Ampersand

In the world he has left for no other place, apartness and togetherness are opposites. At the same time they are inseparable: each requires the other. This is the principle of *between being*. In order for people to be linked, they have first and last to be apart, to be distinct. Conversely, for there to be an explicit distance between entities, they have to be linked. Hence the few yards between his upstairs and his downstairs, the two hundred or so miles between his home and London, gaps to be crossed, spaces traversed by intentions and footsteps and carpets, roads and railways. In the non-place of death, there are no links and no apartness. There is not even the weakest link, the most tenuous bridge: the sunless land is a land without 'and'.

And 'and' – the purest connective – had had protean manifestations in his life. At its very least, it signified mere succession, the link between the present and the previous or between the present and the next. Or the mere, indifferent side-by-sidedness of a blade of grass and a rock, a dollop of birdlime and the shadow cast by a cloud, with nothing in common other than itemhood. 'And' was a tack signifying a passionless marriage, a zeugmatic union between this, that, and the other, linking entities that had in common only their being parts of the same list, for which the entrance qualification was minimal. The scattered were brought together, but as a mere word heap, so that the scattering was not cancelled, even less redeemed.

'And' was forever on the verge of 'and so on', and 'and so forth' and so 'on and on and on', a series that lacked an inner logic, a repetition that did not have the power of rhythm, a succession where each element added nothing but extensity, being mere continuation that summed only to chaos or, short of chaos, to that which in RT's world or his life had taken on the character of 'one (not even damned – for who had time to damn them?) thing after another', occupying one day after another, a jumble of objects, or succession of events. This was the desert where 'et cetera' stretched into the shimmering, thirsty distances, parched of significance.

It is easy to be snooty about 'and', as the pathognomonic sign of mere succession that generates only weariness, *ennui,* or worse. But that was not the whole story. Bringing together disparate items yoked by 'and' sometimes gave him the sense of the great spaces enclosed by the hypersurface of his life. A world that contained 'ships, and shoes, and ceiling wax, and cabbages and kings' was some territory, notwithstanding the hollows in its chambers, and that those elements did not even have the assembled order of a milling crowd.

He had relished a certain subset of 'ands': those signified by the ampersand, by the & which had an heraldic, or aristocratic air: a patrician 'M. de And' or 'Herr von And' that was more than a mere tack or stitch of succession. He had been glad that his had been a world where Dandelion had met Burdock (rarely corked and an excellent finish), and Marks & Spencer and Abercrombie & Fitch did business. The & that joined G with T seemed to trap in its meshes the light of a summer evening as the first drink before dinner, cooled by ice and sharpened by lemon, fizzed with the pleasures to come, that included perhaps Ale & Pie. The Lion & the Unicorn stood on their hind legs to support the shield of the nobility of a country. In the George & Dragon, the ampersand was the lance that connected the dying dragon with its canonized executioner, ensuring the immortality of both, so that they had not passed away, if they had passed towards in the first place. A special favourite, which he had associated with his childhood, was the teaming up of Tate & Lyle to squeeze sweetness out of the world (under conditions he did not like to contemplate). He could think back through the decades to a golden lion depicted on a tin of syrup, the sun-king of animals, couchant in death, being transformed by bees into honey, under the motto 'from the strong shall come forth sweetness'. He would remember the sunlight in the picture on the tin, itself fading in other sunlight in a grocery window, in the 1950s shop owned by Miss Steele who 'went funny' among her many cats. Curds & Whey came from even further back: the dim nursery realm where Little

Miss Muffet and her kind lived courtesy of and-suppressing rhyme. There was the royal & of the V&A and the & that connected the profoundly unequal partnership of Being & Nothingness and the & that carried the expectation, the regulative idea of thought, that to every Q there would be an A. & not infrequently sealed the partnership between great massifs, such as between England & Wales. The supreme amongst these was the enduring ampersand of his adult life: Mr & Mrs RT, joined in secular matrimony, a state of affairs reflected in the millions of envelopes addressed to them both, as they had walked hand in hand for so many years, a hyphenation of lives brought to an end through some A or E.

There were hints of that special connectedness, deserving of the logographic pretzel, in tenuous links he noted between disparate items: in the dimpled appearance of the tarmac on the station platform, replicating the negative pimples on the napkin in the station café, and the waffles he had eaten there, the bedspread smoothed an hour before, the knitted tie he had tied shortly after, and the inverted pebble-dash of the thenar eminence of his hand when it had been pressed on shale sand; or in the mirroring of the dense cobweb of cracks on the white tiles in the wing of a caddis fly, the skeleton of a dead leaf, and the wrinkles on the sun-aged face of a nonagenarian.

He loved similes, underlining a likeness in the unlike. In 'A is like B' two became one while remaining two. The great gap between them was compressed into a thimbleful of sense and yet kept open. His favourite simile for similes were certain elegant motorway bridges, leaps arrested in permanence, over endless hurry. A handful of metaphors, that had asserted the identity of items that had known nothing of each other, had become companions for life. The alcoholic Consul in *Under the Volcano* has his first tequila of the day and he is like a burnt out tree struck by lightning that flowers at all its branches. The despairing colonial officer in *Time for a Tiger*, struggling with documents in the sweltering humidity, cannot stop the sweat, blotting the words he has written on the page, dripping off his nose like a clepsydra measuring out the end of Empire.

They achieved in miniature what those ornate bridges, connecting banks between which great rivers of meaning flow, attempted *in extenso*: the mighty ampersands of works of art that conquer 'and' by embracing it in order to find deeper unity in multiplicity. Thus the arch in architecture and music that arrests time. Or stories that for a while had lifted him above the tyranny of time and space – by linking a beginning with an end, such that each was present in the other. There were connections italicized by rhymes, rhythms, and motifs that iterate sameness through difference.

Ampersands were human-scale intimations of the great ampersand of the world as seen through the icy, disinterested gaze of physics that peered into Nature through the Parmenidean lens of mathematics and saw that the Many were One, a single wave function that refused both to break and not to break.

Coming Apart

In togethering RT had found most of his joy, his anguish, his freedom, his bondage. 'Tween' being, required that he should be distinct, and hence apart, as well as together. Sooner or later, he had known, apartness would get the upper hand and would continue its conquest until he had himself fallen entirely apart and was no longer capable of being either together with or apart from another. It is not love but the laws of nature physics that conquer all, not the least by eradicating the difference between the conquered the conqueror, and thereby emptying the meaning out of conquest. That the dead we mourn do not, cannot, mourn us is the saddest thought of all. RT, after all, cannot even miss himself.

Metaphysics is a way of making some unbearable truths about our lives bearable by turning them into something between tasty puzzles and impenetrable mysteries. So, to lighten the tone, let us dig deeper into the philosophy of togethering and apartness.

Each of us is a viewpoint, a point of consciousness that arranges the world around itself as near and far, here and there (and we have seen how many modes of these there are). We may imagine a universe that, in itself, is a continuum. Physicists do, as we have seen. For them, the basic stuffs and forces out of which the world is composed are not discrete, divided, located. Material objects – pebbles, grass blades, Mr & Mrs RT – are not separated and so do not have to be connected. An extra something that has no place in the material world is therefore necessary to divide it into bodies, events, stories, lives, that have to be separated and connected. That extra something is human self-consciousness.

To be born is to inaugurate a new locality (tiny in the boundlessness of the world) and a new path in a crowded space incessantly scribbled over by billions of feet propelled by the needs, desires, goals, and hopes of a life that wishes to continue to live and to live ever more abundantly. Out of this original contingency we forge our contingent lives which, because they are the only lives we have, seem essential, necessary, and conceal the accidents which made them possible. His being 'I' transformed the contingencies of RT's world into a self-scape in which the course of his life was characterized by 'Of course!'.

The separation of death has exposed the truth of the accidents that brought them together, including the accident of the separation that made connectedness possible. Mr & Mrs RT had been synthesized in separate wombs out of material gathered from the far corners of planet Earth, itself a condensation of boundless space. Their paths had crossed. Their lives became and remained for a while entangled. And then their ways parted. He had been the first to go, to be returned ultimately to the nowhere in particular whence he had been harvested. The million knots of their love dissolved in the great Not of death. Thus their universal story, a tale that signifies nothing – and everything.

Bereavement is the most asymmetrical of all relationships. It might be thought that, as she goes deeper into widowhood, the contingency of their marriage would become more apparent in the divergence of their destinies: she goes to the shops while he dissolves in the rain. They are even further apart than they were before they met, when they did not know each other but at least occupied places on the curved surface of the same earth. But this was not how things turned out. Their love, after all, was a joyful acceptance of this contingency, turning an accidental encounter into the beginning of an essential relationship in which each was necessary for the other. Their connection had acquired an unchallengeable obviousness, resting on an assumption which strengthened over the years that their necessarily temporary cohabitation, in which each was at the centre of each other's lives, would be forever.

While their death-annulled marriage had been the largest and most significant storm of their lives, contingency had ruled throughout their particular, mostly shared, paths: the children they had had, and the details of those children's lives that had preoccupied them; their jobs; their circles or arcs of friends; their houses and possessions; and the many other things to which they had become together, or separately, attached.

Which is not to say that either had started out as an entirely general possibility. Contingency and accident had *framed* their lives. They had lived within unchosen constraints laid down by the time and place (in many senses) into which he and she had been born and by the singular identity arising out of their parental origins. But the possibilities each realized were always only a small part of what might have been possible for them. Despite this larger, and largely invisible, framework of contingency, the rule of accident was largely hidden. There was, after all, the obligation to *be* (or, more precisely) to 'am' a subset of possibilities. The road you are going down is one of many roads in the megalopolis; but it is still the road *you* are going down; and the sum total of the roads you have taken is the only life you will have. A combination of unchosen givens and partly chosen takens, of unsought events and fulfilled intentions, chance encounters and planned meetings, had woven the narratives and ambitions, the responsibilities and rights, that had in turn shaped their shared life.

The quasi-Lucretian consolation for death and widowhood turned on the fact that they had managed to live full lives before they met. He had not missed her in his first twenty-two years when her name had been unknown to him. So they should be able to live without each other after one or the other had gone. Would it not be the case that, as his death receded, the contingency of their years together would become apparent? Although, the habit of being in the same space – of talking, thinking, working, child-rearing, playing, holiday-ing together, weaving a dense network of mutual expectations, such that their lives had become for much of their existence a single fabric – had made each take the other for granted as a child its parents or walkers the earth beneath their feet or breathers the air in their lungs, would not the survivor's years of Not throw all this accidental knot-tedness into relief? As habits of expectation weakened and they were again and again unfulfilled – that noise was not his return, that voice not his presence – time would cut the individual knots one by one. Might not the immeasurable distance between here and nowhere betray how their life together was the accidental meeting of two accidents, that only habit had translated into a necessity?

Even so, loss was loss.

The unbridgeable distances, the apartness that shadowed their years of togetherness, was highlighted as he fell ill. The micro-widowhoods of doing separately things they had hitherto normally done together, of hospital stays or engulfment in symptoms where he was beyond reach, were harbingers. The distances were in plain sight as the bridges between them were broken. The terrors – his of dying and death and hers of his dying and his death – were faced alone.

The tapestry of their togetherness was slowly unpicked. He was pulled away and apart. The fragments of his fragmented self had less and less to do with one another. Plucking randomly at the coverlet seemed to symbolize his attempt to reweave the threads of his life, of himself. Until at last the body that had underpinned the coherence of the world lost its own coherence. Apartness triumphed.

He leaves the world, unaccompanied even by himself, beyond sympathy and antipathy, beyond the warmth of others. He is turned eternally away, an earless deaf ear, an eyeless blindness. It is this absence that populates the rooms, the garden, the road, the village, the city, the holiday places, the pastimes that passed time, hours of the day and the hours of the darkness, and the minds of people they knew together. The years of happily taking each other for granted, filling the world corner to corner, have as their corollary a world which is now evacuated corner to corner. She is Mrs Absence in a universe where his absence is everywhere and yet has its localities; is continuous and suddenly knocks on the door or fills a chair with an RT-shaped vacancy. Those who would comfort her know that they are powerless to do so. Their comfort, after all, is intermittent – patches of kindness, momentary distractions – while her grief is continuous.

The ampersand in Mr & Mrs RT is emeritus, retired on to envelopes, old letters, and other documents of the archive they had gathered. Together.

No river, no banks, no bridge. The end of 'and' is the end without and.

thirteen

Inner Space: On an Extinguished Flame

Inert, dispossessed, voiceless, enisled. And cold: his extinction is palpable to the hand whose squeeze goes unanswered. That modest helping of matter in the bed had, until a short while ago, been warm with mattering. RT has ceased to matter to himself and, sooner than it is comfortable to think, will matter to no one else, either. Tumbleweed, animated by a wind coming from nowhere and going no place, rolls round the empty arrondissements of a personal world where the opening or closing of a door, the giving or withholding of a smile, had been of the utmost importance. In that body, now effortlessly clenched in stillness, there had glowed desires, panic, terror, love, bitter disappointment, slight hurry, mild irritation, quiet satisfaction, faint hope, and a dozen modes of importance and self-importance, the experience of glory and ignominy, a sense of the urgent, the vital, the crucial, the lost cause, and the all-to-play-for and the pointless, the tedious, and the dull.

So, not one but two flames have failed; for the exquisitely regulated metabolism of his body had supported another fire profoundly different from the conflagration burning in stars, in forest fires, and radiating from those many hearths where RT had warmed his hands, different even from the combustion by which those hands, clasped, had warmed each other. It was a flame that had illumined the universe in virtue of allowing its light to become brightness, and making of it an object of knowledge. For the want of a more illuminating term, let us call that inner flame 'his self'. The possessive pronoun betrays an ambivalence: the self is suspended between something he had and something he was; between a possession and the condition of being able to possess.

What went out when that inner, invisible and intangible flame was extinguished?

A Family of Selves

We may set aside the notion, popular among thinkers in RT's lifetime, that his self was an illusion. Such modesty was not only false but self-contradictory because it would require quite a complex self to harbour such an illusion and an even more complex one to argue about it. Only a sophisticated I could argue that 'There is nothing corresponding to what I call "I".'

The (non-illusory) self remembers what he was at a particular time, at different times, and how he was connected over time. Perhaps the greatest mystery of RT's extinguished self was his unity at a given moment, notwithstanding the multiplicity of elements that had gone into his making and, the other side of this, his coherence and constancy, in the face of that multiplicity as moment succeeded moment. An extraordinary, extended, compressed, connected-and-disconnected world was sustained by the existential tautology of his being what he was or (by no means the same thing) being the I that had said 'I am'. We need to advance gingerly as a vortex beckons.

His experiences, his thoughts, and his memories, and some inchoate sense of agency, of being a protagonist in a multitude of small- and large-scale narratives, were all part of the unity of his being RT. So much had to hang together in this future-looking, past-rooted 'now' to make it possible for him to have a sense of 'me' in a world that was generally familiar or, if unfamiliar, unfamiliar in a way that made sense or promised to. He had some grasp of who he was and where he was and where (in the widest sense) he was going to and had come from. Understanding how this could be lay beyond the reach of the most sophisticated philosophy and psychology at the time of his death and looked like remaining so.

And then there was, or had been, the visible and invisible, outer and inner connectedness, binding his together over hours, days, years, and decades. There was one obvious master-link, the necessary condition of his being a person in the moral and legal sense, and his being a citizen: namely this now vacated body, presupposed in all his experiences and actions. It had changed continuously but only slowly – though the pace will henceforth quicken in a dressing down that will unmask his head as a skull as his face becomes a liquifying death mask en route to complete effacement.

This relatively stable body was the basis for the audit trails that linked RT on a day in 1964 (say) and RT on a day in 2012, 17,500 tomorrows-become-yesterdays later. But organic continuity was not enough: it was too close to the continuity of a tree or even a rock. There was also continuity of offices – links in a chain: junior doctor, middle-grade doctor, consultant, etc. – and, more profoundly of relationships – newly-wed, married forty-two years, or being his mother's eighteen-year-old son and her sixty-year-old carer, friend of one year and friend of fifty years. But these, too, seemed predicated on more immediate, intimate continuities. There was a stability of characteristics – kind usually, predictable generally, irritable under some circumstances, risk-averse, cheerful, and competent in some respects, incompetent in others, witty, a less witty recycler of jokes, averagely fond of swearing, responsible and irresponsible, conscientious and careless, a tendency to lay it on a bit thick when it came to both praise and criticism, a voracious reader and constant writer – which influenced the expectations the world had of him and hence, through the pressure of those expectations, stabilized some of his characteristics. And there were the fixtures and fittings of his built world.

These, then, were some of the structures that supported the cocoon maintaining his self; they ranged from habits of thought to the inertness of bricks.

Whenever he looked back over his life so far, RT saw that he had been neither a discontinuous succession of moments or epochs nor a single unwavering flame. He had been a family of overlapping selves. This was, of course, literally true: he had been an infant child, someone's second son and third child, a great-nephew, brother, spouse, father. But in another, more troubling, sense he had been his own ancestor and descendent, a cluster of siblings, rival versions of RT. Each had passed into the other, forever interpenetrating, as he looked backwards and forwards, stepped back from himself or sank into the immediacy of his being.

There had been no clearly defined phases corresponding to the capitalised Childhood, Boyhood, Youth; rather an imperceptibly changing existential flavour (and a changing palate to register that flavour) in parallel with the implicit, often unnoticed, drift in the taste and contents of what counted as his everyday world. Any edges between successive selves were as elusive as the boundaries surrounding a mist or, more precisely, the imaginary boundaries defining the places where fog gives way to mist or mist to clear air.

The backward look was fraught with temptations – most notably an idle confusion of the facts of his case with his past self (of which more presently). Foremost among these was something like 'the condescension of posterity' that saw previous incarnations of RT as raw, his hilarious haircuts and styles of dress evidence that he had, until the wise and mature present, been a half-formed thing, somehow defective compared with the present self in relation to which he had been a mere stretch of *en route*. This Whiggish approach to the history of RT was countered by a dismaying suspicion that he had become dried up, or in an ill-defined way less awake, more closed in or off, stiffer, more of an automaton, less generous-spirited, though also less touchy. The earlier self (whom he scarcely remembers) would not therefore have bowed to the authority of the later self (unanticipated and scarcely recognizable) presuming to speak for both of them, nor accepted that the later word was the last word. Indeed, it would be difficult to know, if each despised the other, who would be the more justified.

RT *père-fils* the historian of RT *fils-père* was to be distrusted for the reason that all historians are to be distrusted: history is written, if not by the victors, at least by the survivors. And how many selves had he survived, or outlived? Every season, not excluding the first, had been marked by death. March daffodils had died, May blossom had become litter, many roses had withered before summer. Beyond the forgotten biographical details, there were lost tones of consciousness and self-consciousness, lost flavours of first-personhood. He would have outgrown and outlived so much. The standard trajectories –

child–boyhood–youth; single man, childless husband; father of young children, father of children entering middle age – left so much behind, with gains offset by, perhaps outweighed by, losses.

What different worlds these allotropes of RT had inhabited! Bodies, appetites, preoccupations, duties, expectations of those around him and the world at large (and they of him), timetables, material circumstances, social positions, futures and pasts – all changed as RT 1964 drifted towards RT 1999 and RT 1999 developed and undeveloped towards his final incarnation. The familiars of his adult life – his wife and his children – were beyond the imagination of his teenage self. Their central and ubiquitous presence was taken for granted though they had not even been envisaged in the twenty years of his coming to his adult self.

Not everything was different of course; but even the more permanent fixtures – those avenues and trees that had survived the wrecking ball and the chain saw – would be differently experienced by successive RTs. Think of the same tree when it was casting its shadow over his pram, when climbed by him as a boy for a long afternoon of reading, as the subject of an incompetent adolescent poem, as rushed past when he was busy and worried or delighted, and as seen from its base when his own child climbed unsteadily on to the first branch. Or think of samples of the conversation with a lifelong friend in 1964, 1981, or 2014, each coming from different places and expecting different things of each other.

Truly to return to an earlier self would be like re-entering a dream after day has dawned, if only because the present self and its world is the arbiter of what counts as wakefulness and reality. For RT 1999 to enter the skin of the boy who eagerly sought top marks, who played conkers, for whom smoking elderberry pith by a fire on the waste land near his house was the cutting edge of vice, who lived in fear of the bullies waiting for him on the way home from school, who had never fallen in love or been preoccupied by sex, and who had only the thinnest buttering of private means, social standing, and curriculum vitae to distance himself from the current account of the day-to-day encounters with others – this would have needed the imagination of a novelist of genius.

When he thought, or tried to think, about his childhood, it served up the same scanty anecdotes and images, much of the material dried by repeated visits to mere husks of fact. Even the 'long afternoon reading among the boughs of a tree' was just a noun-phrase that he could animate only by infusing it with snatched-at images of doubtful provenance. The effort of memory should have been rewarded by an exquisite luminescence derived from the interaction between the inked light in the interstices of the adventures on the page – dappled on the floor of an imaginary jungle or caught on the blade of a scimitar – and the lights round the edges of the second-order adventure of his reading adventures up in a tree. Alas, the backward glance afforded no such flashes. This may have been because of something to which we have already alluded: the value he had placed on his childhood had been undermined by experiencing it as merely a period of *waiting* – to grow up and to live a life of his own.

At any rate, his own childhood had seemed like a sulky, taciturn, shy, or secretive child that refused to engage him in conversation (perhaps on the general grounds that you do not talk to strangers – and no stranger would have been more strange to his boyhood than his late adult self) or to answer his questions. It had nothing to say, for example, about the years in which his mother must have wrapped a towel round him when he came shivering out of the sea, or he impatiently tolerated her rubbing suncream on his little torso while he was pointing to some item in a rock pool, when he was carried because he was too tired to walk, or the years of quarrelling with his brothers and sisters over who had what biscuit or whose turn was first, or entered rooms imagining he was the great cricketer Brian Statham, every carpet edge a popping crease, the lad who found jokes about wet paint and bricks concealed in hats funny and had only to say 'booze', 'swig' and 'spew' to his best friend to cause both parties to be helpless with laughter, or who, in a slightly later carnation, sang along with pop music, looking 'cool' in his own eyes until he encountered the mocking gaze of one of his endlessly mocking siblings. The moments that pressed themselves upon him of their own accord, recalled from the long years (longer for being the earliest years) when he was short-socked, short-fleshed, a bungalow among skyrise adults, and looking up to most faces, were disproportionately those of humiliation because they embarrassed him still, though all witnesses had probably been erased by time or their witnessing wiped by failing memory. And while he could truly remember the joy of making distances his own when he got his first pair of wheels and he was free to cycle round his own city, he could not feel it in the largely fake glimpses afforded to his backward glances. Of the anxious, earnest, little boy, desperate to please, to be helpful, conscientious in his homework, and his transformation to a shy adolescent who retained most of these characteristics but in addition was secretly arrogant, despising the triviality of the world, and the conversations that were to be had in it, and consumed by despair – he retained no coherent sense. Equally beyond reach were the hours, days, weeks, he had spent playing on the wasteland next to his house – the dens 'they' (a term that encompassed a ballet of alliances and quarrels) built, the shed that was the headquarters of a club that had its own magazine, written largely by the editor, the games of cricket, the track and field components of the Olympics, the afternoons in trees.

So little was retained (and even less of that first year of his life when he had so often been cradled on the slopes of a giantess – past to all his pasts), it was as if the intervening years had been a succession of head injuries or a progressive dementia with a retrograde amnesia that wiped away the time when he was dragging on mummy's hand, running a stick along railings, or hoisted on to the shoulders of the dimly, intermittently, patchily recalled father of his pre-school years. The few memories that came unbidden seemed to retreat as soon as he focussed on them. And the predictably meagre results of the endeavour to trigger voluntary memories were soon lost, just as the glass in a framed photograph in the living room, mirroring the present lights and shadows around it, half concealed the moments of the past and made the preoccupations of those squinting at past sunlight and holding hold-it smiles even more remote.

Just how much would elude him of earlier editions of RT had been underlined whenever he tried to capture a very recent past – a moment a mere minute gone. Chasing such moments was like trying to pick up a blob of mercury. The harder he pinched, the faster it shot away. Frustrated, he had sometimes focussed more sharply on the now, and the most purely now of now: bodily sensations, the pressure on his buttocks, the slight pealing of tinnitus between his ears, the faint effort to maintain his vertical posture, the intermittent coolths in his nostrils as he breathed in, the not quite thud of his heart, an itch on his cheek. He had endeavoured to be aware of being aware, at the same time, of the lamp on the table at which he was seated, the table itself, the lights and shades of the room cluttered with so many objects, with a particular square of sunlight on the carpet; more remotely a voice round the corner, someone coughing, a generator sounding like something waiting to move off; more remotely still the interdigitated whooshes of the traffic outside the building. He felt into his own actions: sitting upright; the movement of his fingers over the keyboard and their light tapping as each key sank into its cushioned base after a pause in which his fingers had communicated their dryness, slipperiness or stickiness to each other; and glancing over the letters emerging on the screen. No wonder the past had been so elusive. As for past thoughts…

Memory: A Closer Look

His thoughts! The flame's innermost flame. The head that was once RT's – now unlooking and unseeing, unlistening and unhearing, unsniffing and unsmelling, unsmiling – is (unthinkably to its late owner) unthinking. Because the invisible, inaudible, odourless, cogitation that had accompanied him all his days has ceased in this thoughtless head, its silence was deeper than any he would have known in life. But to separate a scaffolding of thought from the smoke of feelings of sadness, blankness, boredom, tension, joy, loss would be to traduce his consciousness and, *a fortiori*, the faculty of memory that had seemed the guarantor of his temporal depth.

Memory had been the necessary light by which his thought could make sense of itself in virtue of being familiar.. Some of this past would come to him in the form of images, though images were, as he discovered to his frustration, liable to be disconnected. There were feelings that seemed like the bouquet of a past world but, lacking reference, were vulnerable to being 'filled in' with fake facts, by an inner commentator, too eager to explain. Even real facts stood to real memories as dried specimens to the flowers in the field, not truly speaking *of* the past of which they speak. The light of most recent seconds bleached their 23 million or more equally packed, distracted, multi-connected predecessors, when, as in his final days, he had glimpsed more than he saw, and sensed that there was so much more to the world than he could live out or even pass through. And yet his presence to himself had not been a short, dissipating contrail. Despite so much being lost in transit from minute to minute, year to year, decade to decade, he had not felt himself to be a temporal invertebrate.

Memory. It is a luminous mystery undiminished by the disappointments that attend any inquiry into it. While it is true that to live to the full is to grasp one's self and to do so is to remember oneself, the harder RT tried to remember himself, the more elusive he would prove. And there were no reliable summarizing backward glances; no super-self rising above the family of his selves. The last self, in theory best positioned to look back on the sum total, might not even be able to remember breakfast. When all the selves are in, there is no one to look back at them. Or, less gloomily, it is unlikely that one time-slice of RT would be able to marshal all the others and place them side by side in an album in which we see the man in full.

Minor disappointments, surely, to set beside the miracle that had been an everyday occurrence in RT's life. Someone says to him 'Do you remember such and such?' and, setting aside a quibble or two over details, *he does* – or did. Such self-directed, or -commissioned, recall was astonishing. Invited to reminisce about a particular night out, he could glance through intervening space and time – say from his study in 5, Valley Road, Bramhall on 28 September 2014, 8:53 a.m.

to Polzeath Beach circa 11 a.m. 11 August 1995 – and his inner gaze having alighted there, recall as much or as little of that day with a striking concreteness of specification. Such recall required him to trans-pierce hundreds, indeed thousands, of intervening yesterdays to touch the target and light up, and sometimes be scorched by, memories that were not infrequently considerably better than hazy. That is why one of the most striking features of the self – true at least of the instance RT – is that though it went on *so long*, though hour was piled on to hour, again on to again, this did not leave its moment-to-moment entirely dwarfed. The successive nows were not oppressed by the growing hill of beens. It was easier to appreciate the breadth and complexity of the self – that it encompasses recalling facts about Cornwall, geopolitics, medicine (of which more presently), a tendency to be distracted by other people's music, a long-held ambition to be a successful writer, a conscience that was sometimes stronger in feeling than in application, a love of pine trees and the poetry of Philip Larkin, transient worries about itches, spots, aches, sudden pains, more enduring ones about the safety and happiness of his children and his patients and so on – than its persistence. But persistence in the face of growth and change was extraordinary.

Without amnesia – say, for all but a few of his tens of thousands of encounters with patients and the details of those encounters – there would have been no backward glances of any length or depth or distance (all the necessary spatial metaphors are equally inapt). For a teenage experience to have been remembered, the 3,650 days of his forties had to assume an at least temporary transparency, lacking even the faint watery-green of stacked panes of glass. The ruthless efficiency of memory, brushing aside everything that stands in the way of its targets, that attenuates or truncates our past, and our past selves, of the invisible distances we have travelled inside all those visible journeys of shoe leather and CV, is a necessary support for our sense of temporal depth, though we do not experience it fully.

To experience his temporal depth fully, RT sometimes tried to hold together in his mind (say) the nearest yesterday, last week, Christmas just past, the summer holiday (with the beginning and end separated by their 13 fat days), his younger son's twenty-first birthday party, his first week as a research registrar, his wedding day, right back to that moment he recalled (or thought he recalled) when, possessed by an audacity atypical for an excruciatingly shy little boy, he stood up in front of his fellow six-year-olds and joked about the teacher who, unnoticed by him, had returned and was waiting with arms folded beneath her ample bosom for him to finish, looking at him with a glance that imprisoned him in a page of the Thesaurus where 'silly', 'cocky', and even 'cheeky' were inscribed. Or he might try to connect the height-of-summer openness of being hurled by an ocean wave, ridden on a bodyboard, to the sandy shore with the depth-of-winter interiority of groping for a dropped Christmas tree bauble in the attic. Even then, much intervening life would have to be effaced in order that his inner distances could be seen, measured, and felt.

What hope, then, for the resurrection of those perfect hours, imaged in a tabby dozing on a window sill in the sunshine, when sleep and clouds merged on a dove-grey afternoon? What hope, given that 'remembrance of things past' in those 'sweet sessions of silent thought' was anyway disconnected from the exigencies, urgency and demands of the present, against which it seemed a luxury idly indulged in?

Asked to remember himself, he could not make up enough to restore his past, to turn it into a story of sufficient concreteness. Even so, there was one story that seemed to run from near the beginning to near the end of his life. The story of his becoming and being the *someone* who had lived under the name RT.

Enter the Curriculum Vitae

The random, disconnected, patchy character of the past exhumed by memory had been concealed by the notion of the curriculum vitae which carried such authority that obituarists obliged to respond to his demise would turn to this first in their endeavour to tell a wider world who RT had been. Biodata (parents, place and date of birth, siblings, marital state or states, and children), education, career or careers, achievements (including honours and prizes), contributions to the world (if any), 'outside' interests, with a seasoning of 'telling' personal details, were assembled into a compilation that could fit on a page or two.

Add to this résumé entries in various almanacs, encomia, pixels of a picture scattered over documents and archives, and it would seem that the basis of a glance from one end of his life to the other, a coherent portrait, might be available. But the addition sum was never experienced by the sitter or performed by anyone else and so the portrait was only a virtual one. What is more, it was general; not as general as 'His life had seven stages: infancy, boyhood, youth, adulthood, fatherhood, retirement, dwindles' – but still a long way from the flame, from the 'what it was like to be' RT or even what it was like to be with him. He had watched his name make its way through the world as he might observe the progress of a friend or his child or, sometimes, a stranger.

Of course he was many things at any given time: worker (in his case doctor), companion, husband, father, son, bosom pal, casual acquaintance, citizen, fellow passenger, law-abiding queue-joiner; in the same month a wearer of smartish pinstriped suits, white coats, pullovers, and unbecoming shorts; in the same day a wise head leaning forward to attend to a patient's story, a partisan head shouting encouragement at a sports day, and a lover's head, head to head with his wife's head. And that is merely to distribute him between general – that is to say generalizing, that is to say, denaturing – categories. And within these 'roles' (what a simplifying term, borrowed from the stage in which entire lives are presented through a sample of actions and contributions to actions captured largely in words) there was a multitude of sub-roles, in which the lines he spoke were often (how often he could not say) the lines he learned: like his fellow citizens, his speech, where not outright parroting, was for the most part boilerplate. Dr RT might have been remembered as a sympathetic listener, occasionally a pompous ass, a good team worker, a bad team worker, a kind and caring boss, and pain to work for, a layer-on-of-hands, a prescriber of nostrums, a calm and a dry-mouthed panic-stricken presence, an obedient junior, one who did and did not suffer fools gladly, a wise counsel, an embarrassing idiot, idle and hardworking, ambitious and disinterested, conscientious and at times lazy. Who knows? He had been something between Everyone (or everyone in that position) and the particular one that brought his singular history and circumstances and character (whatever that might mean) into play. If he had donned masks, it was less to conceal something that he was, or to posture, but to confer some kind of order or substance on himself.

The official account of that life was almost as remote from the life being lived as blood stiffened in his corpse's arteries was from the warm stuff circulating round his living, running, looking, thinking self. The CV – even with the garnish of 'telling personal details' – was the barest bones stretched over years, a carapace, whereas his life was a flame that flickered and burned in the interstices of moments, the blush in the flesh. It spoke in summaries: 'Between 1986 and 2006 he ran weekly epilepsy clinics.' Thus are caged 1,000 afternoons, perhaps 20,000 encounters with patients; and yet they are all lost, with the exception of a few survivors, the odd face, the occasional crisis. 'He was married for fifty years.' And yet, how much is captured of the 15,000 nights he spent with his wife or (not, of course, the same thing) of the 15,000 she spent with him? The résumé lost the life lived in the middle ground between flickers of inarticulate awareness and 'the shilling life' for the sake of finding a structure, imposed on an existence that had had its unities – but not those kinds of unities – and coherences – but not of that kind – and could no more be collected within it than air or flame grasped in the hands.

While the archive not only bypassed – and so was less than – the life, it was also, in a different fashion, bigger than anything that could be lived. The lived moments were, as the dancing shadows of this year's leaves on a mighty trunk, the living foliage summed as dead wood. Compared with his deposit account, the sum total of his assets, his current account, the flow of revenue, was a minute trickle passing into and out of a lake. Paradoxically, it was the sense that he was unequal to the archive – whether it was in the form of a concise entry in the University Calendar, or his name sprinkled over office doors, envelopes, scientific papers and reports (author, co-author, subject), conference programmes, his own intermittent diaries – that gave the CV a certain authority.

Storytelling

He might have sought consolation in the thought that life cannot be purely the flame that glows in it. After all, not even a fire is all flame, otherwise there would be nothing for it to burn and it would not be tethered to any particular place. To that extent, the CV expressed an existential truth: he could not have been entirely a succession of lived moments. The present had always to be part of something bigger than itself, rising above the succession of grass blades pointing each in a different direction, even if it could never amount to the fake geometry of the lawn.

There were goals, and goals that were part of goals, and goals that were parts of parts of goals. Occupations and preoccupations, projects and ambitions, tied together hours, days, weeks, and even years. The moments transcended themselves in the unities – the commitments, elected by him and imposed on him – of which they were a part. Behind it all, perhaps, was a wish to converge on a single spot; to be a truly integrated being that the necessary virtue of multitasking seemed to subvert; to add up to something. At its heart would be a fulfilment of the romantic dream of *Le temps enchainée par l'amour.* The love of one thing or one person.

To will one thing, to be an unflickering flame, had not, of course, been possible. So he had settled for ideas of self-development, improvement, spiritual enrichment. At the bottom of it all was the notion of a master-narrative connecting a beginning – chosen by the end – and an end – that had been chosen by the beginning.

He had told innumerable stories about himself and his primary, most constant, audience was himself. There were stories of accumulation: pure accumulation as when he counted himself to sleep, each night resuming at the previous night's total, at a land of quantities where few had ventured; less pure accumulation as he had collected exam scores, personal bests in sport, bicycle miles clocked, and circuits of this that or the other completed. There were books, music, places, under the belt *en route* to being something between a cultured person and a know-all who nonetheless thought he wore his erudition lightly. There was, more standardly, the story of his ascent (usual noun) through the ranks (the military metaphor being standard) of his profession, the increasing weight of the bibliography and the CV that had perhaps after all captured something of him.

All these self-centred ambitions bolstered the very notion of the enduring self which they had presupposed. Somewhat less self-centred were the stories of his marriage and children, where his aspirations spread out to encompass others. And, least and most important of all, least and most self-centred, the ambition to understand the world in which he found himself.

As the years totted up, he could increasingly look back on an RT who had not only 'come a long way', but had 'come on a long way', though any eminence seemed large compared only with the nappied Lowlands where his storytelling had begun with a howl. Accumulation gave the appearance of being more than mere addition, though his increased existential girth, reflected in the number of people who directly or indirectly depended on him, the quantity and complexity of projects (with their attendant paperwork, e-work, and voicework) could just as well be seen as fat rather than muscle, or as dissipation rather than augmentation.

If the larger-grain stories – the parallel growth of professional scope, the progress of his children towards independence, the move from one house to another more spacious – seemed to confer some kind of unity or at least fibrous structure on himself, they did so because their coherence did not depend on his continuing consciousness of them. For long stretches of time he could be the sleeping partner of strands of his biography. Because there are stories which are undoubtedly false, it had been easy to exaggerate the quantity, and kind, of truth in stories that are free of actual lies. But this did not alter the fact that what was sufficient to make stories true – or at least not-actually-false – did not amount to the kind of material out of which it would have been possible to weave the tissues of a living, conscious, self-conscious, and self-narrating item like RT. The obituary, while not untrue, was far from the flame.

There were sources of coherence in him, a little nearer to the flame perhaps, that he did not have to narrate, or will, or imagine into existence and which would not have figured in his CV. Beyond the continuity of his body were the stability or inertia of habits, predispositions to action and reaction, a certain tone of presence, of voice, an idiolect in part absorbed, in part cultivated. Naturally, there were also mandated preoccupations, contracted and covenantal responsibilities, implicit and explicit promises, the under-side or private face of the items noted in his CV. These were the non-negotiable connections between RT at time t_1 and RT at time t_2 that had made him accountable for past actions as *his* actions. The price of denying these connections would have been to promote a world in which no one owed anything to anyone else. The fundamental human state of mutual recognition would be put to the torch and relationships would be stretches of exploitative co-habitation without any continuing justified expectation each of the other.

For RT as a node in a network of reciprocal obligation, the connection between his past and his present – or at least certain parts of the past and certain parts of the present – far from being elusive, was inescapable. At the most serious level, RT at any time t_2 felt guilty of mistakes he had made as a doctor at any time t_1, however wide the interval between t_1 and t_2. 'I was young then and I would not make the same mistake now' did not absolve the speaker of responsibility for the suffering he had caused; rather the iteration of 'I' underlined the sameness of identity of the man who had made the mistake and the man who tries to imply that he has moved on.

At a less serious level, perhaps, his being owned by, or at least bound to, his past was evident in the exquisite embarrassment he continued to feel up to the end of his life on behalf of earlier time-slices of RT. Those toes, now upturned forever, had curled, those buttocks, now lax either side of the plugged anus, had clenched at: the episodes of incontinence for fear of asking to go to the toilet, the silly jokes, the crass comments, the groundless boasts, the posturing, the false erudition worn heavily, and the rest. The sense of having been contemptible or at least snigger-worthy was transmitted almost undiminished from the perpetrator of the distant past to the succession of future auditors. There had been times when he had wanted the earth to swallow him up – a wish to be fulfilled, too literally and too late, in a few days' time.

Knowledge

We come to a contested place in the dialectic between the trellis and the vine, the fuel and the flame, between the life that was reported and self-reported and the life that was lived. In common with all those who passed through this place of stillness and insentience, RT had been a place of knowledge. Had he not been a hoard – and a hoarder – of facts, he could not have functioned as a professional, as a father, as a friend, or as someone walking safely down the road.

Whenever he paused to think how much he – and Mrs RT and Everyone Else – knew, he was daunted by the challenge of classifying, never mind itemizing, it. History, geography, literature, art, sport, science, current affairs (to mention a few that occur at random) – each had been copiously stacked in his mental attic. Local and general, active and passive, everyday and specialist knowledge, world affairs and newspaper gossip, vital information and cognitive bric-a-brac had floated and sparkled through his days, motes of dust in his living daylight. The date of the battle of Agincourt, the capital of Iceland, the last novel by Theodore Fontane, the difference between Impressionism and Post-Impressionism, the name of the most prolific test scorer in English cricket, the middle name of T. S. Eliot, the significance of left ventricular ejection fraction, the classification of the epilepsies, the cascade of events following a cerebral infarct, the enzymes involved in the transformation of purine to uric acid, the current prime minister of Australia, the percentage of GDP devoted to overseas aid, the professional name of Silvio Berlusconi's teenage mistress – all could be accessed by his unassisted inner search engine. Pages and keys extended the real and apparent ('Africa in a nutshell') reach of his fingertips.

While his knowledge – like that of everyone else he knew – had been at best an atoll of islands in an ocean of ignorance, it had nevertheless been a legitimate source of astonishment to its owner. Likewise, the disorderly order in which it had been kept in readiness for the moments when it might be needed. And the idiosyncratic mindscapes in which 'the history of English Literature', 'English history', 'England' had been realized in his consciousness. His wobbly, patchy, cognitive map of 'England', for example, was both distorted and true, like the Union Jack in a breeze. He had been ready at any moment to think the spatial relationship between St Ives, Cornwall and St Ives, Bucks, to make an informed guess at the population of London, to recall the entrance to the Queen's Hotel in Leeds, to bring to mind Chosen Hill, Caer Caradoc, and the Malverns, and to list as many English composers as any audience could stand.

These general facts were supplemented by vital local knowledge he seemed effortlessly – his brain spared the rack – to retain and access as required: the charges in the car park near his favourite pub, the details of the final chapter of a thesis he has advised on, what to do after he had turned left at the traffic lights en route to a dozen destinations, the addresses of many friends, the precise spot where a conversation with someone a couple of years ago had left off, the whereabouts of the keys to the sliding doors opening on to the terrace from the kitchen.

On the borderline between direct and factual memory, any reminiscence session would have been able to turn up reliable recollections (as well as unreliable ones of course) of, for example, how the light had fallen across the floor one evening ten years ago, the tone of voice in which someone had said something during a brief conversation (one of many millions he has had) in his youth, the taste of a beer he hadn't drunk for decades. There were memories that counted as knowledge, and a kind of achievement, that had to be memorized, and memories that came with experience, hardly an achievement, requiring no rehearsal. This was a distinction that psychologists would have captured in the difference between semantic memory (the date of the Battle of Agincourt) and episodic memory (the location of that conversation with a friend). The two often crossed over – as when he remembered learning about the French Revolution and remembered also the way the light fell in the 1950s classroom as the brilliant Mr Roxburgh dictated his notes about the guillotine, allowing his prosthetic arms (the real ones having been blown off by a grenade) to crash down to the desk to illustrate the arrival of the beheading blade and to wake up certain students who were rather drowsy at 3:30 p.m.

The facts were timeless and the experience of learning them had been located in RT's inner time. As for those facts, even those about his own past seemed to belong to an aerial view that sees whole towns but misses what is in the interstices, such as the looks of love or anger exchanged between two people in the bedroom of 27, Acacia Avenue. And past experiences of learning the facts, or episodic memories, had not seemed to carry the aura of temporal distance. They lacked the mould, or moss, or must, or the aroma of a past time, which they should have had.

The pursuit of the inner flame, the vine within the trellis of The Life, needs to take another direction.

Inwardness

'The self' is a word we use to designate many things, among them an inner space which, not being like smoke in a box, is a mode of the inwardness of space that is explicit time. We sometimes forget this and think of that inwardness as most clearly manifested in things he kept hidden, in the privileged knowledge he had of himself. And there is further a tendency to think of the hidden as, most importantly, that which has been kept *secret* and RT's corpse a safe that could not be cracked because the key and the code have evaporated with his death.

Of course, there were things that he had kept secret, including images that would have awoken his now forever-inert member, now stiffening not through a rush of pulsing lifeblood but through participating in the processes that are making his entire body a stiff. Any sins, griefs, angers, even joys that he might have felt ashamed of were now beyond his judgement. And they had, anyway, formed only a minute part of the hidden that is now buried forever in the shortly to be buried corpse. His deepest secrets were not ones that he kept concealed but ones that he would not have known how to reveal.

The most important obstacle to self-revelation was the banal and yet fundamental fact – upon which we have already brooded – that he had always been himself, uninterruptedly in his own presence, and unable to escape awareness of what *he* was experiencing and feeling, where he was, what he was up to, and what was preoccupying him. The encounter with others was always intermittent and those others were to a lesser or greater extent opaque. He had to infer what it was like with them, their experiences, emotions, intentions, and preoccupations, from signs that they might emit involuntarily or voluntarily. Any communication would always be incomplete – even where there was a commitment to openness – and subordinated to a variety of chosen and imposed constraints. And when they were not in his presence, his knowledge of what, where, who, and how would be sketchy at best. A chance-met man with the lank comb-over, nodding in agreement with everything he said, and saying 'Yes' at the end of every other sentence, was just an example of a universal truth – the opacity of others – that should be more obvious.

The other barrier, almost as formidable, was that he could not have directly compared and contrasted his own experience with that of others. There was the inescapable mediation of language which generalized everyone, and their experiences, in order that what was communicated should be intelligible to all parties, erasing what might be the essential difference. Admittedly, the language through which each self shared itself with other selves would already have coloured, shaped, and structured those selves – most evidently in the way they talked to themselves – so that less was lost in translation than might be the case had each self been presented to itself innocently of any prior verbal contamination. Even so, the gap between what it had been like to be RT – how it had felt to be him from moment to moment and how those moments connected up into a minute, an afternoon, a term, a period of his life – and what RT said about RT was huge, or at least immeasurable.

It would be a different kind of mistake to exaggerate RT's hiddenness and think of his self being some kind of private, stable self-image, concealed by the thick veil of the meat that was now his corpse. After all, such glimpses as he had had of himself when he walked upright in the daylight were mediated through the idea of what others thought of him, and through the language that was available to him – the idiolect he had carved out of the shared tongue. It is this that makes even self-ascribed comparisons of character traits seem questionable. Secondly, imaging himself was always prompted by those particular occasions when he was prompted to, often painful, self-reflection. And finally, as soon as he allowed his thoughts to drift to something more interesting, or when the world made demands on him, or he was caught up in some action, what little fragments of a self-image he had formed would be shattered or simply fade, albeit not as completely as had happened during the painful journey of disintegration that had ended in this place, where he is lying still under a sheet.

There had been moments when he had projected himself outside himself and observed that projected self in action. These Out-of-RT

Experiences had seemed perfectly ordinary but they had been none-theless miraculous. Take one example, picked at random from a population of billions. He arrived one morning at the gym and discovered he had left his headphones behind. The miracle of his body responding to his efforts with a precisely judged tachycardia and tachypnoea, sweat (controlled as to composition and quantity) water-wrapping his head and gathering to runnels, was not sufficient that day to stave off the boredom of the exercises necessary to post-pone death. And so he constructed an imaginary stadium – made out of what fabric? – in which he would run the self-prescribed kilo-metres on the treadmill. He had then populated the awaiting track with his running self. Of what fabric had that self been made? I rest my case.

Such arias of explicit self-representation were intermittent. Although RT knew enough, for example, to scheme and plot or even innocently and benignly plan what was to his or another's apparent advantage, it would be a romantic simplification to think that some-where between that grey beard and that bald head there had been a kind of surface in which RT was mirrored and he saw himself, and even his true interests, clearly; that what he saw left nothing to the opacity, impersonality, and timelessness of factual knowledge. While he had had a privileged (if often incomplete) information about where he was at a particular time, what he was doing, and most importantly (because invisible to others) what he was feeling, thinking, desiring, planning – not to speak of all the silly things he said to himself and the daft ideas going through his head – this was offset by his inability to get outside himself. Which is not to say that others' views were necessarily more faithful to the truth about him than his own; for they, also, were mediated by their own situation, the customs and language through which they were used to make judgements or feel responses, through their own limited knowledge of themselves, and of course the limitations in their capacity or even inclination to imagine what it was like to be RT. While he could seem to give a satisfactory explanation of many things about himself – for example, that his choices were shaped by an original desire to please or not to disappoint others and a subsequently to free himself from within the cage built for him by these choices – this was a long way from what it felt like being the explanandum of that explanation.

In Pursuit of Depth

Being RT had been the lived mystery of being always not-quite-here, looking to the future and looking back; regretting, hoping, dreading, waiting, moving forward with purpose and resolve (not as sharply defined as the words that referred to them), drifting and dozing, waking and sleeping; creating a rich present out of an endless dialogue with a personal and shared past and future. His ability to continue an action, to remain true to a purpose, in a sea of qualia, holding on to things that demanded attention and unpeeling from those that had to be ignored, in a network of spontaneous trains of thought and effortful thinking, involuntary memories, and the children of racked brains, had been astonishing. It had prompted RT to try to grasp something corresponding to the idea of his substance by finding links between disparate memories, thereby getting a measure of the distances within himself, and making his objective temporal extensity explicit as depth. Like history, RT had many cunning passages, wormholes that linked days that knew little or nothing of each other.

He had appealed to the four years of his progress from the Orange

Volume 1 of Whitmarsh's French course to the final Purple Volume 4, via the blue and green volumes, each of which marked an epoch in his boyhood. How fresh the volumes seemed at the autumnal beginning of the school year when they had been distributed by the book monitor on the instruction of the teacher. And how spine-wrinkled, dog-eared, blotted, scuff-covered they looked as the summer term came to an end. And how exotic, advanced, Volume 4 seemed compared with introductory Volume 1.

This yielded less temporal depth than he might have hoped because the hundreds of morning and afternoon lessons and evenings of homework inserted between maths and history – indeed, everything apart from the gathering scuff marks and inkblots on the textbooks – were lost to memory. He looked to another example: the interval, already referred to, between his first hearing the phrase – designating the noisy upper middles who annually invaded the little Cornish seaside village of Rock, referred to as Kensington on sea – 'Hooray Henrys' used by his friend and a moment many years later when he employed it himself for the first time. Had this been a means of linking, via a secret passage, the moment of a friendship, in the summer life of the village, a particular evening, the light in a bar, with the occasion several years later when the phrase popped out in conversation with a friend of his now grown-up son in an urban pub in December?

Yes and no. It remained as difficult to be (say) the temporal depth of one who was sixty years old as to live the long life one has had: time does not add up. Stacked-up days collapse like a card castle on itself. The sixty years lay largely beyond his mental and imaginative reach. They could no more be recovered than could the wax of the Advent candle – gathered back like honey – be reconstituted out of the flame that had silently recorded the passage of December, even though the flame had given the purest sense of the passing days, and its reflection in the eyes of his excited children the most numinous image of a Coming, of the Christmas glow getting ever closer in the darkest days of the year. He endeavoured to excavate the depth of the past in the sudden image of a face across a room, or a starry Mediterranean night or a child, at peace at last, sucking its thumb, of the tunes he found himself singing, the moment of regret, the shared joke that made him smile again, a phrase associated with someone who had once been important to him. He kept samples of the Post-it notes which for a decade or two had been key to the organization of his life. There was poignancy in these messages he had sent to waylay himself with reminders of things to be done, these memoranda-turned-memorabilia, addressed to a future self now far in the past. Or, again bringing space to the aid of time, he would recall the gleams in the glasses in a *Mitteleuropa* café, metonymically capturing the light from a paradigmatic city square, where he had watched the comings and goings above the edge of a book he had read and now largely forgotten. Or he imagined himself in old age, waking at 3 a.m., thinking back to games on the beach with his young children in the sunlight. Or recruit the coloratura blackbird's glistening song to connect the half-asleep child, kept awake by the long evening light of summer, with the other half of his sleep in old age.

These fragments of private antiquity, however, did not order themselves into a structure that would have held open the depths his lived time ought to have become. Nor had extensity stood proxy for temporal depth. It was true that the growing mass of his possessions and the density of connections forged at different times were proof that he was more than a succession of days. His thousand books, his hundreds of CDs, his photo albums, his family, acquaintances, friends, correspondents, should have stood for the existential co-presence of all the eras, facets, and places of his life. And while he could read, and think, only one page at a time, it was a fat book he was reading and he could sit next to shelves of books he had read or intended to read. But present moment still had the power to elbow aside the rest of its life with its unchallengeable seriousness.

He had tried other ways of revealing himself to himself. He would reconstruct the interval between the beginning and end of a Greek holiday a decade before, as seen from the last day, before the return to work had squashed the entire fortnight to the unprivileged periphery of his consciousness. Or he would imagine a moment from that holiday – say, a glass of retsina on the sharp-shadowed terrace of a taverna overlooking the sea, perfumed by pine trees wired with cicadas – being recalled from the nook of a shadowy, rain-swept English pub, cashing both aspects of space-time into time. He would connect the moments when his mother had gently rubbed calamine lotion on to his sunburnt neck with a vision of his own son, a tanned toddler, asleep after a day on the beach, and the same son overlooking the same beach from a headland he has just climbed with him, defending Michel Foucault. Or recall a lunch with his other son, and walking away with a sense of loss, after a long reminiscence about that young man's childhood, a reminiscence itself now in the distant past. He had imagined a kind of redemption of lost time in the multiplication of past tenses, perfection of memory in pluperfection, though 'had', 'had had', and 'had had had' sometimes clattered like a growing pile of slates.

Internal Stitching

He had been distracted, multiple, and in some respects incoherent. At a certain level he was a mere succession of experiences, feelings, thoughts, memories, roles. Nonetheless, 'RT' was not merely a legal, or any other kind of, fiction based on the relative stability and unity of a piece of meat bespectacled in its later years. This we have established. The flame within him had a constancy that transcended the flickering of his moments. He was fastened together in ways unknown to this corpse, whose arm was attached to the shoulder and the head to the shoulders, unknown even to his body when it had been a going concern. That which indubitably connected the frightened little boy entering the large, noisy spaces of the reception class overseen by a lady who was not his mother, to the just-retired elder whose education and its fruits were largely complete, driving past a noisy primary school on the first day of term, had made it possible for him to glance across space and time to invisible places at long-past times.

He could say 'Lambeth Hospital' and evoke an image, belonging to a world nearly half a century before, and 2,000 miles away from the screen on which this sentence is appearing, of the fussy, anxious, small Irish Casualty sister who famously overlooked the fact that the girl with abdominal pain was giving birth. Or to imagine how his study had looked before its latest rearrangement. Or to look out of a library window and see the storks flying over the thick woods, overhung by clouds and illuminated by lightning, behind the little Janáček museum in Hukvaldy which he had visited on a sultry July day.

This ability to reach through space and time – such strange distances, such instantaneity, and such precision – had been a striking reminder of how much of him had been in place; of the extraordinary coherence of these past worlds waiting to be illuminated by the appropriate prompts. As for reminiscence, saying 'Do you remember that day when...?' or 'the look on the face of the man with the pink shirt...?' or 'the dress I was wearing on...?' was not only a bridge-building, or bridge-affirming, mode of connection and a reassurance of a public reality underpinning private memories, but also a shared metaphysical miracle.

He had fluttered round these capacities, both moth and flame, conscious that they had seemed ordinary only because they were shared with everyone he had known. If he had relished what he had called his 'internal stitching', it was not because his stitches were any more remarkable than the other modes of his connectedness but because they seemed to make the strength of that connectedness even more extraordinary.

RT had sometimes found it surprising how much (common, ordinary) sense he had made of himself, of his world, of his fellows, and of his life; how he had managed to surf the disorder, the fever of mere association, the silliness, the flotsam and jetsam of thought, to maintain an inner as well as an outer RT who had shared so much intelligibility. Which was not to deny that had again and fallen away from the chosen self, the self of 'us', of the role, the CV, and biography. Self-command, and self-supervision, had been dissipated by local distraction and global doze; until at last he had reached this final, furthest, irreversible fall to a place beyond pulling himself together, coming to, to a non-place where the work of remembering RT would be to be carried out by others. All his life – from the time when he drifted to sleep, thumb in mouth, reciting the same silly little phrase to himself, to nodding off over his homework, or an important briefing paper, to the afternoon dozes of his later years – he had been subject to occulations: gaining and losing himself, becoming absorbed in local things and waking up to the bigger picture and, briefly to an I, a who-what-when-where-why, that was not confined to small parish of am. Involuntary switches from Beethoven's Ninth Symphony to an itch commanding complete attention, absences of a mind that was present to times and places that were remote from the matter at hand, mistings over the mirror of consciousness: these were as much The Story of RT as the tale told in the entry to *Who's Who* or in the obituary added to the verbal necropolis of Monk's Roll where the lives of dead Fellows of the Royal College of Physicians were assembled.

Farewell

That elusive, inescapable flame had taken on many hues that would have not figured in the most comprehensive biography. The dapples in the patterns of philias and phobias, of sympathies and antipathies, of affinities and repulsions, would be beyond the tweezers of language. Much of the available vocabulary seemed to wither in the inner spaces we have been talking about. When, if ever, was he 'in cahoots'? Was he 'a stalwart'? What side of the border separating regular from irregular guys did he stand? How close was he to being a 'scamp' or a 'holy terror' when he was a smallster, or a pompous ass as an oldster? Where did he lie on the scale between 'fusspot' and 'negligent'? The lexical tweezers would have proven even more rubbery when put to determining whether or not he was good at judging – in kitchen or Intensive Care Unit – whether or not things were tickety-boo or how 'swish' were the places in which he located himself. What pen could capture the ways in which he had been irritated by a biro: its running out, its refusal to write on inhospitable surfaces or at particular angle, its incontinence – blotting, running, leaking in pockets and on shirts – and, most egregiously, falling down the lining of his overcoat and catching his left leg at each step as he had strode with purposeful stride to some purpose? Or do justice to the theory, the justification, and expression of the minor joys and aggravations creating their distinctive hues in the flame? Or the modes of deliciousness – air, food, smells and other sensations, music, and gossip – he experienced?

Thus the inner flame, RT's self, a mysterious continuity beneath the pell-mell of actions and events, feelings, experiences, memories and thoughts, a silence at the heart of the outer noise and inner chatter, has guttered for the last time.

Snuffed.

It.

Towards 'The Late RT'

fourteen

Auguries of Insignificance

Naturally, he had seen this coming. The general principle was clear, though he had not foreseen the path death had beaten towards him. He was aware, after all, how small a figure he cut in the sum total of persons, places, and things, and his consequent utter dispensability.

He had often attempted to internalize the great spaces in which he was cast in order to put some distance between himself and those compelling concerns he tried to experience as 'petty'. Since he was of a secular cast of mind, it was not the imaginary gaze of God (Whose infinitude would make his own finite portion of being infinitely small) that conferred miserable wormhood (or less) upon him. It was science, with its numbers, that exposed his smattering of flesh and his smidgeon of life as insignificant.

He had known since boyhood that his likely lifespan would be less than $1/200,000,000$ of the age of the universe at the time of his birth. And that the interval between the tips of his outstretched arms occupied one 001.84 kilometres out of the $42,000,000,000,000,000,000,000$-kilometre width of the All. Even if his reference to the earth as 'his' planet had been literally true – as if he not only dwelt on it, leaving a few tiny scratch marks, but encircled it with ownership – this would not have made him signifi-cantly larger. For the earth was a minute particle in the space swept out by the solar system and the latter was but one of 200 billion galaxies of other planetary systems housing 10^{22} planets. The great auditorium filled with light was less than a pinprick of brightness, the city in which he passed his childhood was a wormcast, and his nose was absurdly close to the ground.

These daunting figures, however, had rarely cut existential ice. His aspiration when young to cosmic laughter had been at best posturing. And far from being shrivelled by the sense of his own insignificance when he raised a numerate gaze to the sky, the fact that he knew and remembered the statistics that quantified that insig-nificance had elevated his estimate of himself. It was, after all, only in virtue of minds like his that the universe was a whole – was 'the universe' – and nothing else in the 93,000,000-mile interval between his head and the sun warming it had the slightest inkling of this great distance. Besides, the numbers were rehearsed only by permissions granted when he was in a glade of comparative peace – painless, tranquil, unhurried, and unworried. A glance of disapproval from a fellow microbe could occlude the gigantic numbers that should have revealed him to be of nanoscopic size. While, being an inveterate counter since early childhood, he was inclined to grant that the quan-titative had more authority than the qualitative, it proved impossible sincerely to hold that neutrinos were more real than backache or the beauty of the beloved's face.

The reflection that he was invisible from what would count as a microscopic distance in planetary terms did not marginalize the fear of an upcoming examination, or divest any of the commonplace embarrassments of their burning significance. Very little space was left for the Milky Way, with its billions of stars, by the thought that he had made an idiot of himself. A smirk on another's face would be enough to extinguish the message told by distant starts. In the end, the RT-diminishing thoughts were brief and insignificant episodes in the life of one who was, according to them, brief and insignificant.

Admittedly he occasionally entertained the ghost of the Pascalian conceit that he was somehow superior to the universe, which vastly outsized him and would crush him, because he knew the universe while the latter did not know itself. He was confident of belonging to the ontological top drawer, notwithstanding his difficulty in locating himself in the sum total of everything. But of course, in this respect, he was not alone. And therein lay the rub. He had belonged to a current of humanity that had swollen to a flow of 7,000,000,000 people, the present representatives of a primate species that over hundreds of thousands of years had walked the surface of the earth in pursuit of food, drink, mates, health, and ever more complex notions of happiness. The evidence of his insignificance had consequently been set out more compellingly in the trousered, frocked, loin-clothed biomass of crowds visible and invisible than in the statistics attached to the material world of which his body had been a material part.

Numbers sometimes still played a part in awakening him to his real standing; as, for example, when his son mentioned that the Metro in his favourite city, Prague, had supported 500,000,000 passenger journeys a year. There was no reason to assume that these journeys were any less rich in purpose than his own. All that business – those trips and meetings, and that heartache, boredom, fatigue, worry, dailiness, transported over the city – had been as real and significant as his own. The report that in 2014, 423,000 students had taken up places at UK universities put his all-engrossing three-year undergraduate adventure into momentary perspective. Contemplating the number of his renewed passport – 512189689 – threatened another intermission in any sense of being special. A short-lived pause, of course, however hard he tried to imagine the faces, journeys, and lives behind the numbers.

There were more frequent occasions when his being a mere One amid so Many had impressed itself upon him more directly. The milling, swarming, jostling hordes among whom he milled, swarmed, and jostled, literally pressed the truth of his condition upon him; namely that he was vastly outnumbered by individuals to whom he did not matter. The size of the All made him a negligible part of the sundry. Nevertheless, such large numbers of people, and their overwhelming consensus that his concerns were of no concern, that his dry-mouthed hurry was unimportant, did not crush him.

How could this have been so? How could he have withstood being so small, so few, and so transient compared with the crowd – so large, so many, and seemingly eternal – thronging his days? Because he could turn the Many–One asymmetry on its head. It had been *his* viewpoint that had gathered up all those others into a crowd: *he* was one and they were atoms of 'the many'; those who had outnumbered him were halfway to being mere numbers, counted by him. Moreover, the 'hordes' who surrounded him were rarely a threat: they were not as visibly endless as in India or Mecca where each was separated from the edge of the crowd by acres of human biomass threatening death by trampling underfoot or suffocation.

Elevated, on the balcony of the Britannia bar at Euston, above the pushing and shoving of one embodied subject against another, and enjoying a drink before boarding the Pendolino, with its huge ingestive capacity (a fat worm absorbing hundreds of organisms through its side-pores) RT was the invisible eye gathering up his fellow passengers into *his* crowd crowding the crowded station. Likewise, when his gaze had shrunk the flow of cars on a distant motorway to a procession of aphids, his sense of being a mere unit or atom in a traffic jam of other individuals whose needs and intentions got in the way of his needs and intentions, was greatly mitigated. Thus the illusory supremacy of the unwatched watcher over those he watched.

And so he had walked through his days for the most part undiminished in his own eyes by being a nameless extra, or less than an extra, in the lives of the many millions of extras who had shared actual or imagined or simply rumoured spaces with him, negotiating worlds woven out of thickets of intentions that were not his. He could therefore withstand seeing that he was 1/50,000th of attendees at a match, or knowing that he was one of a readership of 2,000,000, or one of 10,000,000 or 500,000,000 viewers. Those others were after all intermittent whereas he was continuous; episodes – for the most part minor episodes or (in the case of members of a crowd) small parts of minor episodes – in his life. The intentions of these transients were not, as were his intentions, wired into a week, a lifetime of connected meaning. That hurrying person over there had been merely a stretch of hurry whereas RT's hurry over here was a thread in a much bigger story, interwoven with many other stories into a dense fabric of significance. What was more, he had some responsibility for whatever it was that required him to hurry. This felt responsibility – however unwelcome – underwrote a self-importance not seemingly justified in the man in front of him who had happened to be blocking his descent into the Underground by standing on the wrong side of the escalator. Those other bodies pushing and shoving against his had

not been visibly populated with narratives, inwardly lit with imperatives. While he was the permanent protagonist of the salient dramas,
often interconnected, of his seventy-five years, all but a few others
had had only brief, walk-on, or shove-and-push-on, parts. His non-
negotiable self-importance had made the self-importance of others,
however at times irritating, on the whole easier to bear because marginal. It was not an existential threat.

Besides, the crowds eventually dissipated as he proceeded to the
private, crowd-free spaces that awaited him at the end of his journey.
He returned to a street, a house, a room, and to conversations in
small places where he was large, not outnumbered, and his tongue
was more than a single fluttering leaf in a forest of voices. Home
('at last'), with a glass in his hand, surrounded by his own people
and by things that reflected him back to himself, he could declare
that London had been crowded, the station had been crowded, the
train had been crowded, and the crowds had been awful. Because
he could not hear or even imagine the other members of those
crowds, also returned to their own spaces, saying pretty well the
same thing, locating him in a blur of unremembered faces, he did
not question his promotion from $1/1,000$ of visible humanity to say
½ or ⅓. He was only faintly aware of his membership of The Great
British Public, each of whom also had a rather dim view of The
Great British Public.

Most importantly, his mattering to himself had now been confirmed by others to whom he mattered and who in their turn had
mattered to him. In the sequestered places, the 5-foot-10-inch midget
RT, in reality dwarfed by trees, buildings, mountains and planets, a
mere bubble in a seething horde, a grain in a sandstorm of faces,
an inlet of the billions who were his contemporaries, had presence,

even stature and importance. The Great Outside was a mere rim on the margins of his awareness. He was spared the realization of basic truths such as that he was a small and intermittently present item even within the envelope of his own house, smaller in his street, smaller still in his city, and an infinitesimally small dot in the nation which had granted him passport number 512189689.

His smallness was also obscured by the story of a life marked by progressive enlargement. He was a giant compared with the child he had been. This was a more salient comparator than that offered by the thought that he was one person and was not 7,000,000,000 other persons. And his standing in the world at large was of less moment than his presence in his children's lives (as theirs in his). He was calibrated against little platoons of others who were equally insignificant, upon whom his health, happiness, self-esteem, and so many other things depended, as theirs had depended on him. This had obscured his ontological lightness of weight, revealed more truly in his being a once-met stranger whose name had been forgotten, encountered between naps on an express train, or a link in an endless chain of forebears and the as yet unborn.

His minute standing in the greater schemes of things (that did not, for the most part seem to scheme at all, least of all against himself) had been tolerable, even invisible, ultimately because his consciousness favoured the small over the large: he was more attuned to registering loving facial expressions and hurtful moments of impatience than to the Great World that accommodated so many millions like him. Indeed, he could feel more completely crushed by an adverse comment on something he had said than by the 50,000 faces cobbling the terraces at a football match whose indifference should have reminded him that he was not merely one out of the tens of thousands in the stadium but out of millions in the metropolis where the stadium was located.

Objective, quantitative knowledge had a superficial hold on him compared with the love, and need to be loved, that had been the very soil in which he had his being. The fact that the universe cared nothing for him for would be itself nothing compared with the fear that Mrs RT might care nothing for him. And her demise would empty the universe more completely than any statistic. Which was why the house in which he lived with nearest and dearest had seemed 'an everywhere' notwithstanding that, compared with the solar system of which he was a part, it was much less than a hair's breadth to the width of the Atlantic Ocean.

If the cup of his significance had been full when it should have been empty, it was because it was small enough to be filled by minor, local significances – irritation with someone who is chattering, or fussing over minor details (a particular bugbear), or over-exercised by logistics (an even bigger bugbear). There was usually enough significance to make his actual insignificance self-sufficiently significant. Tutored by a necessarily exaggerated sense of his own importance, and pacified by a strange indifference to the contingency of the life he had been given, he had learned to tolerate being one of an unending succession of crowds that formed and dispersed and re-formed and dissolved. And a good thing, too; because he had been among crowds to the end of his days.

He had been insignificant even by certain human standards. The world had never found it necessary to burn him in effigy, to offer him police protection against a crush of admirers, or to name prizes, buildings, or cities after him. The places where he was obscure far outnumbered those in which his name carried any meaning. He was for the most part mere noise in others' cursory glances, any old Tom, Dick or Harry, hardly a Rag in the trio that included Tag and Bobtail. The obscurity that was the lot of the overwhelming majority of those who had lived was his lot.

There was one crowd that grew more numerous throughout his life: those whom he had outlived. The roll-call of strangers, the torrent of the named and nameless, soliciting contributions from a rather limited store of sorrow, was unending. 'Man dies in canal horror', 'Elderly cyclist killed by out of control lorry', 'Six children perish in house blaze', 'Suicide bomber slays fifty more', 'Ferry sinks with 100 passengers', 'We remember the 200 of our colleagues who have died since…', 'Outbreak toll reaches two thousand', 'War claims its 10,000th victim': the reports were relentless; the gravestones endless ('shovelling them in by the cartload every day' as Dublin's Mr Bloom so daintily put it); the pyramid of ashes growing without pause.

He had shared his deathday with a crowd of over a quarter of a million: a hundred in a massacre here, a thousand in an epidemic or a storm there, so many killed on the roads, or losing what is usually called 'a battle' against this or that condition though it was a fight they had neither started nor wanted and was more intimate than any civil war or internecine strife. That there had been nothing specifically *his* about his deathday was the ultimate justification of that sense of his own insignificance that he attempted, without success or even sincerity, to profess. For most of those who remained on the planet his death was not even an anecdote or occasion for the shake of a head: 'RT? Was he the chap who…? Or was that…?' In the eyes of the world from which they have departed, the dear departed are usually cheap.

Now, in his death-garb, an inattentive stillness, it was too late for RT to regret how little he had realized the scale of the world in which he had passed his days. And that, though he had cared, he had not cared more. His sense of the insignificance of things had extended so much more easily to others – those millions of strangers who shared the air he breathed – than to himself. There was something in ordinary humanity which had permitted the perpetration of the enormities of his and previous ages, and he had shared it. Obliged by his existing obligations to be indifferent to others' suffering – 'There is only so much you can do' – he was, even at his most generous, savagely restricted in his sympathies and in the scope of what he accepted as his responsibilities. That part of the world that had legitimate expectations of him also fostered his capacity to hold as cheap the lives of the overwhelming majority beyond its boundaries. The tunnel vision that had enabled him to progress through the world in a coherent way was matched by tunnel sympathies.

He had sometimes (rarely) had the thought that it was only a small step from this spontaneous, respectable neglect – occasionally seasoned by a self-justifying anger that wished ill on others who obstructed his path – to a principled devaluation of others. The wider indifference necessary to deliver on his responsibilities – a higher selfishness – or to confer narrative meaning on his life (even the narrative of one who had wanted 'to leave the world a better place') could, he had thought, warp into a hatred, energized at times by the clash of sharp elbows.

It was merely a piece of historical luck that he had not translated a neccessary holding-cheap of the life of most strangers into devaluation of their very existence, that he had not colluded in some collective outrage. When in 1832 the Duc de Rovigo ordered a whole tribe to be put to the sword, while he was at a dinner party gambling and drinking, he was at one with RT in this respect: the lives of others – including their sorrows and screams – had not merely been offstage but off-world.

And now he has reaped his just reward.

fifteen
Outliving

It was as a teenager that RT had been most acutely aware of the death to which he was sentenced. Paradoxically, end of childhood, boyhood, and youth had been more charged with forebodings of the coming darkness than the subsequent anniversaries that marked the passage from middle age to what he knew would be his final phase. This may have been because, as he had approached adulthood, life had become more timetabled and the timetables stretched further into the future. The unformatted boundlessness of the 'not-yet' of childhood had narrowed and started to look like an arc connecting his forgotten, unimaginable beginning with his all-too-imaginable end. This new awareness had expressed itself in many ways: fear, yes; and a disheartening sense of the pointlessness of a life which ended with the obliteration of all meaning.

But there were other, gentler, intuitions of mortality: the sense of the irreversibility of even the most trivial partings, and the absolute loss of a day, or a season, that had come to an end. The sepals shed by dehiscent buds, the browning of blossoms, the hasty gallop from snowdrops to daffodils and from forsythia to hawthorn, from lilacs to laburnum and (summer's first wrinkle) the lime trees flowering and dripping sugar on the bonnets of cars parked under them, were more powerful mementos of transience and the approach of the all-engulfing blank than the withering of autumn leaves and their pruning aided by the invisible secateurs of frost. When spring passed its high tide and the solstice was reached, he had felt a sense of gathering loss. The intervening decades had not diminished the memory of the sometimes delicious melancholy of the summer evenings of his teenaged years.

The longueurs of dusk, its shadows elongated by obliquities of sunlight, smoothed the ripples and the wrinkles in the fabric of the day, evening the odds and oddities of time. Notwithstanding the birdsong enthusing the unfenced heights threshed by the tallest trees, these were the hours when daylight thickened to stifled night, and windows changed to mirrors gazing inward into rooms grown cameral; when the bewitched and bewildered driver, squinting between light and twilight, resolved ambiguity in a fatal smack; when crepuscular shadows shepherded cruelty inwards from the street, hidden from the judgement of the just; when children heard their heartbeats as spurts of blood, encroaching footsteps by the bedroom door.

And yet his youthful self had loved as well as feared the sadness of a July evening, with its breezes that stroked the still silky softness of the leaves, the slight silvering of the flanks of the trees, and their woken voices seemed to massage his thoughts. These were the hours when the suburban roads and streets his recently outgrown childhood self had pedalled through seemed to widen, deepen, and lengthen to tree-lined *avenidas*, coordinates of honeyed combs of ingled inskirts; when walls were softened with tiredness, mossed with human time, and pavements returned intact the kindness of those who cared.

After the bright arterial mornings and the weary venous afternoons, the world in the evening seemed an earful of murmurs, homebound traffic was a rumour of another shore, as if the summed-up day were a wave breaking on a beach as wide as 'now'. Growing hush had invited an attention that listened into glades of quiet for pianissimo notes: the seedlings of Perhaps, paraesthesiae of possibility.

The version of evening that was cupped in houses, glimpsed from the solitude of his adolescent promenades, was of nests of hope, worlds scoped by candlelight, the toy behind the chair, laughter in the cot, the hand on the shoulder, frail words, and glances gleaming in the gloom. But behind this there was always another evening uttered in the hollowed, outsider's moonlight of the barn owl's call, sad laughter at closing time, sounding the vastness between the city and the hills that could be seen from it; articulating spaces between crags and stars; between those stars and yet more stars.

That part of him which had sensed something fearful as well as lovely about evening knew that its coming signified the beginning of outliving, that he was the survivor of another day that some others had not survived, and that the darkness towards which it so beautifully subsided would sooner or later be dawnless. Each day had brought him one day closer to the not-yet-appointed day.

This trajectory was no more direct than his metaphorical journey from forgotten birth to invisible death. His life, like any other, had been marked by cycles: the days cycling from dawn to dusk and back to dusk again; the seasons fostering and stripping off the leaves, husbanding and scything the plants; the months circling from January to January. Within these cycles were smaller cycles – breathing in and breathing out, the heart clenching and unclenching, appetite coming, going and returning. And larger cycles, as the solar system kept to its course. Cycles within the cells of his body, cycles in the weather, cycles all the way up to the great diastole of the Big Bang and the systole of the Big Crunch.

There had been a seemingly endless succession of fresh, or at any rate new, starts: he had been a junior, a beginner, of sorts, at every stage in his life. But the rhythm of departure and return had not restored things to precisely what they had been. The cycles of damage and repair, injury and recovery, of displacement and return, concealed irreversible change, a path from development to decay; something not quite corrected, progression to a point of no return. Those who live by the grace of chance, packets of order in a universe that cannot even judge itself as chaotic, must accept the inevitable.

The time of his death was not written in the universe: it was not special, any more than his 'now' or his 'no longer' or 'not yet' had a place in the sum total of things. But there was an inevitability about the drying up of new starts. Endings began to dominate over beginnings. If death did not ambush him from without, it was written in the biological fatwa that decreed it from within, running towards him a little faster than any flight he could muster. And there was gathering evidence of new beginnings in lives that would outlast his own: the toddler taking its first tottering steps, a barely controlled stumble; the child, shy and panic-stricken like him, entering a reception class; the adolescent astonished at his first kiss. The world would continue with its exquisite attention to detail beyond the moment he breathed his last, the indifferent earth circulating the indifferent sun, the children falling out in the street, the anxious schoolboy hurrying to complete his homework, the lover wild with expectation, as RT in his sick room chased the spectres presiding over his descent into nothingness.

He had outlived so much – so many selves and so many others: he had experienced the end of relationships, associations, and circles of friends; the ends of eras, defined by places in which he had lived, offices he had held, the periods of his children's lives. The entanglement of life and death, of living, outliving, and dying had preoccupied him, who has now been outlived. There had been a subtle withering, drying, and obnubilation as he had come to look more like withered ancestors than fresh-faced descendants. So many selves – with their distinctive fears, joys, preoccupations, and attunements – had faded; and he suspected that he had become coarsened, thickened, blunted, dulled equally by repetition and by variety. He had wondered at times which would despise the other more if at sixty he had met with the fifteen-year-old self that had died into him.

He had been fortunate in having to withstand few losses, though outliving others had begun early. His grandparents, who had not lived to see him, had always been dead and so had never died. They were words, stories and a handful of poorly printed photographs that scarcely awoke his curiosity. Great-aunts were early casualties. Aunt Nell, preserved in a splash of 1950s kitchen light, was remembered untying her pinafore, grizzled of chin saying 'I believe so' and then later in a psychiatric hospital, demented, swaddled in sheets, in one of two long parallel rows of bedfast in-patients, beyond words. She died offstage.

The first real death was that of a little girl, five-year-old Pauline Burns in his Reception class, who ran across the road outside his school into the path of an oncoming lorry. He remembered the little girl in a purple gingham uniform dress mainly because of his embarrassment. A short time after the school was informed of this tragedy, he was overheard repeating the old chant 'Pauline Burns, burns', a joke based on a functional shift of a word between a proper name and a verb. He had forgotten that she was no longer, and was reproved by a shocked teacher. She lived on in his lifelong mortification. Three of his teenage friends died – two by their own hand and one, a boy-racer, in a head-on collision. After this early flurry, there was a pause, leaving aside the patients he could not save. Then the pace quickened to a steady flow as colleagues, distant acquaintances, someone he might once have exchanged a knowing glance with at a meeting, seniors, heroes, adversaries, completed the one-way pilgrimage. As he got older, more casual 'cheerios' were final than he was aware of. And throughout there was a steady torrent of strangers' deaths reported on the radio, the television, and in the newspapers. Casualties of war and peace, they were frequently so numerous that the obituaries of the deceased were often aggregated to mere numbers reporting the totals to which they contributed one life.

And then his father and mother had died at the ages of ninety-two and ninety-four respectively. He had been by their bedsides at the end, able to hold their hands, though he could not give them the comfort they had once given him, could not make them safe from what they feared most, as they had made him safe. His cries, when he was separated from Mummy – when he had found a terrifying darkness beneath the blankets, or in a cupboard that had locked itself, at the back of the garden shed, in the shops when he had wandered off into an ocean of alien faces – had been answered in a way that he could not answer hers, as she found herself cast adrift. As they had slipped away, they were no longer separable from the sea in which they drowned. The promise 'I will make sure you are all right' must always eventually be broken because the universe was, first and last, careless of its progeny and careless of all care.

Those deaths more than any other had required of him that he should learn the art of outliving; a seemingly small part of the art of living but it was a special challenge to learn how to remember and not to be paralyzed by memory; to know the size of another's death without being diminished by it; to feel the loss but not to be eaten from within by loss; to pay tribute to the past without mortgaging the future. There was no right path between the commemoration of mourning and 'moving on', giving the dead their due, their life in his own life, and yet being able to continue the ordinary, necessarily local, even trivial, patterns of thought and action that his life had required. Life would be impossible if the survivors did not forget the dead and living were reduced to outliving; but it would hideously shallow if they forgot them too quickly or too completely. For the most part, those who had outlived him – visited him in his extremis, watched him drift out of reach, even wept at the pity of his death – would slowly forget him, as the memory of all that they had said and done together would be diluted by an influx of fresh life he would know nothing of.

So he himself had survived, as most do, the death of his parents. More than survive: he had continued to prosper and to live a life that had been rich by any standard. Even so, he had hoped that he would go before Mrs RT; selfishly preferring that she should be the one who carried a burden of bereavement that, he had felt, he could not lay down without abandoning himself. This loss would be the place where the elegiac sense of ending – with the last daffodil of spring as poignant as the last rose of summer, farewell to playmates called in from the street to bed and solitude as retirement from a much-loved job – would come to its absolute, its feared climax. An ending where the idea of a new beginning would seem to be an insult to all that had gone before, if only because it suggested that all that mattering had amounted to nothing. True grief does not want to get over itself because the one who grieves does not want to get over himself.

He had anticipated what widowerhood might be like. He turns to her to smile and her face is not there to receive it. He thinks of something he must tell her but she has nothing to hear it by. He comes home to their home and every space – behind the front door, in the kitchen, in the photograph of a holiday home, in the bathroom, in the bedroom in every cupboard where her things are – is hung with a Not that evacuates meaning from what he sees, hears, says, thinks, owns. Her death, a stubborn absence, an unending silence, an eternally averted gaze, an infinite sulk, is everywhere, in everything, in every-one he meets in whose eyes he is, above all, a widower. It no longer matters whether he walks to the right or the left because all places are equidistant from the non-existent place where she is still alive.

The philosophical comforts are comfortless. 'Do not attach your-self to created beings as they or you will lose everything.' Would he have had it that no one cared especially for anyone else? That each, careless of his own life, should be careless of others' lives? A world without unreasonable love, without the special meaning he had in another's eyes and they in his – however little objective legitimacy it had – would be a world without meaning. And a universe under-pinned by a God who (notwithstanding the evidence) cares for the fall of a sparrow as much as for RT, whose love is evenly distributed over His creation (and the current 7,000,000,000 needy souls) is as comfortless as one in which no one is cared for. RT had needed to be specially (that is to say over-) valued by someone who had uniquely valued him for his life to be bearable; for him not to evaporate into a succession of moments.

Looked at from death, his life course (in particular the assemblage of people who had loomed large in it) may have seemed contingent; but this contingency was all he was. He began as something (a human being) and a someone (this human being) in particular; and he had remained something in particular. He had lived his once and forever life within limits. Limits that had become more apparent as the shadows gathered.

sixteen

Limits

Less limited than many, perhaps than most: the item under the sheet, after all, was an old man's body. RT had died in late adult life – even if 'lateness' was a term that could be applied only with difficulty to the sum total of events that was a life, or straddle the honeycombed here of his days and the nowhere beyond them. Though he could not have anticipated the exact date, time and mode of his death, he had reached the age at which the balance of probability between living on and popping off, always open to revision, had started to tilt irreversibly in favour of the latter.

RT had been scrawny, never knowing what it was like to go through the world muscular or fat, but his muscles had still wasted a little over the years while his abdomen had arrived at the first trimester of kegnancy. An acute eye would have noted the shrinkage of the belly of the first dorsal interosseus, between the base of his index finger and his thumb. More readily visible was the time-weeding of his cranium and the little solar keratoses on the dome, the fruit of long exposure to the daylight in which he had lived and, in particular, the sunlight in which he had flourished.

Of more fundamental significance perhaps had been the slight gynaecomastia, the shrunken scrotum, and the thinned grey pubic hair. These betrayed the depletion of testosterone in his body – ably assisted by the enzymes induced in his liver by a steady flow of alcoholic drinks primarily addressed to his brain but calling at most other organs on the way. And more directly symbolic were the deep wrinkles on his face, incised by habits of facial expression, and the more subtle photo-ageing of his face and forearms and the backs of his hands. The skin had slackened a little and the slackness been taken up into a network of folds.

With age he had lost some of his singularity and become 'Old Man' or 'Elderly Gentleman' in the eyes of strangers – a sign of the de-differentiations to come. Pruned by the encroachments of winter, he had found the world a less habitable and usable place: the fire sinking, the air more nipped and less friendly, his legs more aware of their own weight as if the earth had been sucking at his feet, gape-mouthed in anticipation. More of his world had been hidden from sight, round corners he could not reach, silences sat less deliciously on his blunted hearing, his grip on things weaker.

When the expansion that had characterized his childhood and adult life had begun to reverse, the taking back did not mirror the giving. While everything he had built had eventually been unbuilt, and all the messages that had emerged from the noise of time had ultimately been scrambled, he had not simply returned to his first state via the route by which he had come, shrinking to a clump of cells. The path taken by the stuff of which he was composed in its return to the mindless order of the world had not visited the places it had passed through on his way to being fully formed RT.

While the gathering probability of his non-existence had been well signalled, he had no way of knowing which times were The Last

Time. For a start, there was always a remaining life expectancy and in his final decades this had grown as he had aged. For each day you live, he had learned, you increase your life expectancy by six hours. This may have promised a Zeno-type situation in which the hare of death would not catch up with the tortoise of the living man. But such promises were empty. After all, if Zeno had shown how things really were, then the material of which he had made could not have converged on the same spot such that he came into being. And of course the increased life expectancy was an increase in average life expectancy, based on actual data about actual deaths. So mortality could not be escaped, howsoever it might be postponed. So there were Last Times, though their portentous status was hidden from him.

Approaching the end was not like the final year of his professional life, when he could tick off the landmarks – the last departmental Christmas dinner, the last PhD student undergoing a viva, the last epilepsy clinic, the last teaching session, the last patient. There had been nothing comparable to signal that this was the last time he would tie his shoelaces, watch the suds slither down a knife he was inspecting for cleanliness after he had washed it, be embarrassed by an involuntary release of wind, take off a particular or any shirt, attend a reunion and struggle to name faces mussed by the Second Law of Thermodynamics, feel a shiver down the spine, queue for a ticket, pop his head out of the door and declare that it is quite mild, watch frail pillows of outbreathed steam disappear into the dark, buy a pair of reading glasses from Poundland, put down and lose a pair of reading glasses from Poundland, enjoy a sneeze or be irritated by another's protracted bout of sneezing, point out that his surname had two ells in the middle, confuse a patch of light for a peanut or a stranger for a friend, observe the moon untarnishing as its ascent up the sky meant a thinning lens of atmosphere between its surface and RT's eye, swim underwater, make a can of underarm deodorant sneeze a perfumed cloud, watch the warm wind of an oncoming tube train ruffle a girl's hair, say 'See you later' and mean it, fill the car with petrol, scratch an itch between his toes, nip out for a breath of fresh air or a swift half, use the definite article, clink glasses and say 'Cheers', declare 'I love you'. He would have kissed goodbye to spaces and places never to be revisited, to people, never to be seen again, to cherished books to be opened no more, to concepts he had taken pride in using, or names he had pronounced with a sense of achievement, but being unaware of the eternal interruption, had not offered them a kiss or the least ceremonious farewell. He could, however, say with some certainty that 'This suit will see me out' and imagine a world in which all his clothes were unoccupied by his body which had been sheltered and made presentable and decent by them.

As his singularity faced its ending, he became increasingly aware of these and other limits. The sum total that would be RT – all that he had experienced, loved, owned, for good or ill done – became more clearly defined. He had done this, rather than that; gone there rather than stayed here; kept quiet rather than spoken out or vice versa; campaigned for this and not for that; become this expert and not that expert. To the limits ordained by the accidental parameters dictated by his date and place of birth – this century, this culture, this social class – making his life necessarily a small parish of consciousness within boundless coordinates, he had added limits in part elective. He had settled for certain ways of being, for being average or less than average in pretty well everything he had done, for customary modes of behaviour, a particular weekly or yearly round.

This self-shaping, self-confinement, self-entrenchment even, was as it should have been, as each had to fit into a life shared with and supported by many others. It was nevertheless the most profound concession to the principle that to grow up was to be self-pruning. This was the part he played in bringing it about that, in the if-light of possibility, he was a bounded creature in a boundless world, a small buzz of actuality in a great space of the might-have-been. He had chosen to be that small, bounded creature, recognized that horizons that promised limitlessness were outer definition of the limited creature they encircled.

As the shadows began to gather, 'un' increasingly haunted him. He had felt himself squeezed and ever closer to the state in which we see him now, in which he is unable to move, futureless, his past scattered through a world whose rumour can be heard in the traffic outside this room dominated by his stillness, traffic unheard by him. He had become unphilosophically aware of the thousands of roads taken compared with the millions or billions not. The countries not visited, the friendships not made or not kept up, and, nearer to hand, the CDs on his shelves awaiting in vain to be heard, the classic books that he had foregone reading as he had idly reread a lesser novel, or leafed through a magazine he had despised, the bookmark left forever between pages 22 and 23.

At times he had regretted the part his own choices had played in making the sum total of RT smaller than it might have been before he had been kidnapped by the Great Nothing, taken back by the very forces that had brought him into being. An inveterate classifier, he might have classified his regrets under a variety of categories. Here are some of the larger ones: the undone, the unknown, and the unlived.

Let us begin with the *undone*, which might seem the most obvious source of regrets to haunt our corpse in the bed, had it been awake to the life that had ended with so much unfinished. There are some undones that he might not have fretted over too much. A lover of modest-sized mountains, he had never conquered Mount Kilimanjaro, but then he had judged that the effort of getting to the bottom of and climbing its slopes would outweigh any exaltation on arrival at the peak. The idea of the logistics of travel and of sore feet or aching legs or panting chest were potent disincentives. And the Taj Mahal by moonlight (ditto on logistics, with perhaps some galloping gut-rot thrown in) had been preferred as a reference point and a source of jokes than a visited reality. He had known his limitations as a traveller, a looker, a seeker after experience for its own sake. In keeping with his averageness, he had not been unadventurous but nor had he been he particularly adventurous either: while his motto had not been 'live gingerly' he had not aspired to hang-gliding. A short period of rock-climbing had been followed by long years in which abstaining from playing games with gravity seemed wise. He had frequently inverted George Lowe's famous response (when asked why he climbed such perilous mountains): 'Because they're there', with the equally famous joke: 'I like sitting in front of warm fires at home because they're there.' If he had had any regrets it was at not doing enough of Doing Pretty Well Nothing – reading, writing, thinking and undertaking modest Lakeland walks or bingeing on the elemental qualia (light, heat, sea) of a Greek holiday.

It will be evident, therefore, that while timidity may have had a part in some of his self-imposed limitations, equally powerful was a special kind of indolence that resisted plans that involved so much organization, arrangements, and discussion of arrangements, as a result of which the point of the event was rather dwarfed by the infrastructure that made it possible, and the actual doing was only the tip of a pyramid of preparation. It seemed as much a distraction from his idea of experience as a means of achieving it. Fuss, details, and a certain kind of planning were anathema to him. If CBA – 'Couldn't be arsed' – almost qualified as a watchword it was because so much had to be done to make certain kinds of doing possible. Purpose dissipated through so many steps and stages was diluted. Details were devilish distractions, curdling his consciousness to a lower level of awareness from which the Big, Important, Fundamental realities would not be visible.

So the regrets under this heading would not have been particularly clamant. His bucket list consisted mainly of a wish to do more of the same. There are other regrets he would have liked to believe that he would have: the apologies he had not made, the kindnesses he had passed up on, the good fights he had not fought, the moral idleness that prevented him from being stung by empathy into helping when he simply hurried past, his occasional professional incompetence which might conceivably have meant that he had outlived some patients for longer than he ought, his insufficiently continuous sense of the truly important, though his desire to keep the Big Picture and the Grand Project in mind had made him somewhat inclined to solitude, a little precious about time, and unwelcoming of interruptions. These, more than the failure to learn Spanish, how to waltz or play a musical instrument, or the art of fly-fishing, would have been the source of torments any afterlife he might have had.

Besides, one of his ruling ideas – though it did not limit him as much as it might or should have done – was that everything needed for a full life lay pretty close to hand (so long as hunger, illness, fear, anxiety, and grief were in abeyance) and the challenge was not to live more lavishly but to make sure the life one had was not unlived. Looking, thinking, savouring, appreciating, wandering locally about, might for the most time be sufficient – though novelty was also needed, if only to restore novelty to the familiar – the real point of travel to foreign places being to rediscover the home grounds through eyes made a little bit foreign. He had felt that, if bounded by a nutshell – or one anyway with en suite toilet, a bookshelf or two, and a view through an upper-storey window, and space for his wife, children and a few choice friends – he could have imagined himself a king of infinite space. There was no need to see everything there was to see if those things he did see were treated as exemplars to provoke thought. This was, after all, the ontological vision at the heart of philosophy that had a slight preference for Being over beings, and instances as exemplifying types. Likewise, the pressure to add to the sum total of his possessions could have been contained if he had acquired the art of possessing those things he had. This was the spirit that had animated his love for (at the time of his death) his twenty-year-old Volvo on account of the mileage he had experienced in it. And his thought that he would not have to travel very far if he imagined and thought about the roads he did travel. Who, after all, could exhaust a suburban road seen through intelligent, thoughtful, imaginative eyes? So the visit to Peru that he had never contemplated, and even the (now) eternal postponement of the walk down a side-road parallel to the one he had driven past for twenty-five years on his way to work, like the failure to read a poem in the original Mandarin, counted as little as his never having said 'in a nutshell' without at least the ghost of inverted commas.

Life had required of him that he adopt a very particular angle on the world, observe and engage with it from a specific (rather low) altitude, in pursuit of a narrow succession of aims, duties, schemes, plans, and dreams, all those things that made demands on him in response to the demands he made on them. And this was a source of regret. The small-scale and close-up occluded the vision afforded transiently to RT striding over the hills, the struggle into a pair of trousers or to reach a glass of water without spilling it or to focus on improving the running of his outpatient clinic, had blocked the sunlit landscape, the irritation caused by someone's sniffing next to him a few days before he is due to deliver what he described as a 'major' lecture, had distracted him from thinking at the level he had aspired to. This had obscured the miraculous network of networks of meanings and actions that had comprised the most ordinary day of his and others' lives. No wonder that, at times, life had seemed 'the same old, same old' and its length had even seemed a long littleness.

Enough of these obvious regrets. There is perhaps more to be astonished at in regrets he did *not* have, such as the ease with which he tolerated his own ignorance. He had, as we have already noted, lived eye-deep in the *unknown*. Just how much that vacant head had known was extraordinary – as we have again already noted – but how much he had not known and how easily he had tolerated his many modes of ignorance was at least as extraordinary.

Ignorance, as we have seen, began close at hand, in all those places where he had found himself in company, only guessing how he was experienced by them, not knowing how they were experiencing themselves. He had seen the world from many angles but these were always his own angle, the angle he was aware of but somehow could not make fully apparent to himself. The flow of opaque faces that had passed him revealed little or nothing of the history, the propensities, the characters, preoccupations and thoughts of their owners. An exchange of glances was at times akin to a glimpse of lights twinkling in a distant village. The things people said to him had occasionally – when he was most awake – seemed to come from a country of which he knew little. Deciphering what motivated people, how they hung together, what it was like to be them, had been like trying to deduce the nature of an embroidery from the loose threads trailing from the back. Even so, he had steered his head through tides of faces usually without major incident; which may have been why he had settled for allocating people to labelled boxes, assisted by uneducated guesswork.

This intimate, continuous ignorance had opened on to another, as ubiquitous but a little more remote. It had begun in the dappled pattern of revelation and concealment in his sensory field; his ignorance of what was under a stone, round a corner, beyond earshot, out of smell or out of taste, in the next room or in a distant capital. These epistemological limitations had metaphysical roots. Objects of knowledge – even stones and tables – exceed what is given in the experience of them. That is what underwrites their reality. With knowledge, therefore, had come the possibility of error and ignorance, a possibility that extended to the very item he most identified with: his body. The meat beneath his own skin had been an area of darkness which had had as its black moon his lifelong ignorance of the length of his period of office as the going concern 'RT'.

His map of the world, therefore, had been largely of *terra incognita*. He had known little of history, the names, habits, and lives of plants; animals were largely a blank; and his grasp of the implications of the oblate shape of the spheroid of the earth was limited. More generally, the mathematical portrait of the world – the most comprehensive and profound gaze humanity had trained on the reality it found itself – lay beyond his elementary numeracy, so that he had had little idea of the workings of many of the things that he needed in everyday life. The laptop on which he had written so many words was an impenetrable black box. He had loved literature and music but had read only a minute proportion of the novels, poems, essays, histories, treatises, that should have spoken to him – his 'bookishness' meant more '-ishness' than 'book' – and the overwhelming majority of the music of the overwhelming majority of the great classical composers whose names he revered had never reached his ears. And the catalogue of paintings and sculptures unseen by him was vast.

Even in the medical science whose practice had occupied the greater part of his adult waking consciousness, what he had known was as a grain of sand in a beach of what he did not. He had not read all the key articles in his field; in his sub-field, or in his sub-sub-field; nor even the abstracts or key words; or the titles; not even all the bibliographies. That would, perhaps, have been unnecessary as the key facts, sifted by scholars of the sub-sub-fields would rise to the surface and be available in the places where he had needed them. Any shame at his ignorance was palliated by thinking, by way of comparison, of the epistemic darkness in which scholars in the humanities plied their trade. In those disciplines, many key facts were not agreed on, and individual approaches were not transcended in the path to some kind of objective truth. Literature scholars knew little of philosophy; English literature scholars knew little of French literature; Victorian literature experts were ignorant of the views of their colleagues in sixteenth-century literature; Browning scholars had little time for the work of Tennysonians; and so on. New historicists despised post-modernists, Lacanians had no time for neuro-literary critics and so on. The image of hundreds of thousands of scholars writing for a handful of readers, who would read their produce only in pursuit of a completeness of the bibliography of their own pretty-well-un-read works, reminded him of his mother's description (innocent of its political incorrectness) of an Irish parliament: 'everybody talking and nobody listening'.

To think thus had made him feel slightly better about how little knowledge there was in his gaps. And he had sometimes entertained the illusion that he was 'erudite', 'well informed', 'well read'. In part this was because he was able to make a brushing, third-hand acquain-tance seem like first-hand knowledge. There was much he could have said about *The Tanglewood Tales* which he would die before reading. It was sufficient to be on the edge of knowing what a rawl plug did. The theory and discourses of gravy and sauces largely passed him by but he knew what he liked and roughly what went with what, even when he could not reliably name either the first what or the second. The distances that separated cognitive bric-a-brac that littered his mind contributed to the impression of great territories of the known: that the tonnage of meat exported from Argentina increased from 26,000 in 1900 to 400,000 in 1916; that worm infestation in dogs was transmitted by eating snails and that global warming had moved the vectors north; that (since we are speaking of worms) the prospectus of the proposed National Museum of the Czech Nation published on 15 April 1818 specified a comprehensive collection of Moravian and Bohemian earthworms among its contents; and (a favourite of his) that A. E. Housman had spent several decades editing a Latin author (name forgotten) that he thought little of.

He had died, in short, like Every Man, a pretty-well-know-nothing but felt (like Most Man) confident that he knew quite a lot of what was to be known.

The undone, the unknown, and – most painfully – the unlived. 'Live all you can, it's a mistake not to,' Lambert Strether had concluded at the end of Henry James's, *The Ambassadors*. Strether was not of course advocating a life of sex, drugs, and rock 'n' roll but rather one in which timidity and caution would not lead to one's being merely an ambassador sent across the Atlantic to rein in a mildly errant cousin rather than following one's heart, venturing a little to win something.

RT's regrets would focus not on the paths not taken but the taken path not experienced. On the absent-mindedness, sleepiness, indifference, insensitivity, that marked his awareness of the passage – one among many billions – he had taken through the world. Unliving had not infrequently been the product of a habit of politeness, rooted in embarrassment. So his late concern had been not with the life 'over there' that had been denied him but the life 'here and now' that he had been insufficiently awake to. It was not the relationships he had not had, the people he had not known or met, but what had been left unsaid in those relationships he had had, to the people he had known, those he had met. How shyness, distraction, impatience, occasional competitiveness, anger and silliness that surprised even himself, laziness, the desire to be alone, had allowed awkward silence or empty nattering to preclude memorable speech that might have gathered the world in great handfuls or reached into the depths of the person before him. Mutual presence had too often been dissipated in gassing, as each colluded with the other in finding the easiest things to say. Had this, perhaps, been the most grievous cause of the unliving of the life now ended?

This might have been his epitaph: 'He had not been sufficiently present in the world that had presented itself to him'. If he had startled those who had come to mourn his death with pronouncements from the grave, it would have been what every corpse might perhaps say if it could speak:

I wasted time and now doth time waste me.

He had tried to imagine just how time would waste him, once his time had run out. The spider, for whom his left nostril was just another uphill slope, pitching a fragment of a veil, populated with its dead prey, over his face. Crows, with voices like the charred paper from the bonfire of his days, pecking absently at his eyes. His skull, a hollow bust, with an unpeeled cranium, air-filled orbits, his nose, after a million rounds with the fist of nature, broken beyond pugilistic possibilities, and an eternal uninterpretable emoticon-smile, the emptiest of empty grins, revealing the teeth of a meat-eater, fixed in the hollowest of hollow laughs, a vessel filled with rubble, lowlife, and excrement. Animals grazing and flopping on ground where his flesh had given the grass a special savour.

He could foresee this and had endeavoured to keep it in mind; and yet he had still allowed himself to be indifferent to an April evening, glistening with dew and birdsong that could have become itself in his consciousness. He had not tried hard enough to think into the lives and minds of others – the stranger in the street, the child next door, the patients who had come before him – to imagine sufficiently into the sorrows, joys, and inner organization even of some who were close to him. Knowing what was coming, and how soon, he had asked himself, how could he have indulged for a second in the time-vandalism represented by mean-spiritedness, resentment, and bickering with the world?

Enough: point made; it is too late for sermons; and, besides, the art of living well had not been without its irresolvable contradictions. How many mornings had he (honourably) foregone playing with his children in the sunshine in order to work at some project, some distant oeuvre or pursue some dream of changing the world (and of course his prospects in it) for the better? Would it have ever been possible to contribute to the larger needs of the world without being absent from the present, from the here and nows? To frame the question in this way is to exclude the already noted role that Couldn't Be Arsed, a fatal apathy, indifference, or hopelessness, had played in his life.

The very idea of a life fully experienced had we been the equal of it, had we paid undivided attention to it, was, he had acknowledged, in some respects illusory. Time lost can seem privileged only because we are looking at it from the outside, from a present self, a present life, to a past self, a past life, which is fixed in amber. At the time they were experienced, the past hours had been swept by their own immediate past, propelling them to a rather local or narrow future making its demands on him. Now, remembered from a future whose pressures he could no longer re-construct, those distant hours are lifted misleadingly from the flow of events and from the obligation to act. The past is merely pictured, its storms mere patterns of colour on a living room wall, rather than life-threatening irruptions of cold and wet. And so he seemed to have squandered the hours whose possibilities he had not lived up to. But he could not, perhaps, have done otherwise.

Perhaps the deepest reason why the life he had lived may have seemed to have been in part unlived lay in the nature of time. 'The lived moment' is a *movement*, it has momentum; it comes to itself by moving away from itself. His being had consisted in crumbling into becoming. Or (if this seems too deep for the hurry of Tuesday, the woes of Wednesday, or whatever it was – idleness probably – that in part eclipsed the sun of Sunday) losing himself in the moment, even with the aim of opening up to a world wider than his habits of perception would normally permit, would be to lose the necessarily general, outwardly connected, meaning of the contents of the moment itself. To be a human being was to be a creature not of the moment but of the hour, day, year, even the century, of his private history, and of the many histories he had shared with others. How could RT have been mindful of Cornwall when he entirely lost himself in an old wall steeped in sunlight that seemed to take on its ancientness? To lose the moment for the Big Picture was necessarily to be not-quite-there. The existential moment was not the mathematical instant and yet the elusiveness of the mathematical instant hollowed out the existential moment.

So while he may have regretted the life unlived, it could never have been fully lived. There was always something more important than the honey of Now, even though the latter alone was where importance could take root. To live had been to be at once fastened helplessly to the facts of RT's existences and to find that they had eluded him. The hope of a perfected attention to his own life, in which his consciousness would dilate to the idea of a great totality without the loss of darling particulars, remained only a hope because it was self-contradictory.

It might have been expected therefore that, accepting this, he would have been resigned to his limits, and concluded that, give or take a few minor additions and subtractions, and settled for a rather modest sum total of what he might achieve, know, and live. He would abandon the search for happiness, worthiness, or even completeness, forgo the alibi of the future, and cease striving to augment the CV. And declare that here, or hereabouts, it would be the right time to die.

Is there such a moment? When it would right to say, like Simeon: 'I have lived my life. *Ich habe genug.*'

It makes sense of course to choose death if we are in unbearable pain and the weeks or days that remain offer only the intolerable suffering of a decaying organism closed off from the world. But otherwise? Are there not new stories starting all the time and much unfinished business? There is, after all, no point at which all the stories come to a natural end, especially since completed stories are only parts of larger stories that are also incomplete. So long as there is meaning, there is an opening to a future, itself open and awaiting resolution. There is always space for a new project to be started and seen to completion, something else to discover or to experience or to be known, and, most importantly, for more togethering, with strangers, friends, and loved ones, most notably Mrs RT and their joint progeny. RT's world had been continuously finding new vistas: the future of his children; the way science and knowledge might advance; the tantalizing possibility of a more complete or at least deeper understanding of his own nature.

The argument for indefinite life became more compelling as the most subtle and continuous assault on his love of life – the attack from within – had seemed, as he had got older, to be gradually called off. The obvious, the ordinary, the already-known – and that taken-for-grantedness that had sometimes spread into a suspicion of a wider meaninglessness, as if the truest progeny of time were must and mildew – had capitulated before his widening sense of wonder. Those states of mind belonged to the past where irritability, disengagement, and boredom converged. In the past he had known days of seemingly endless repetition, of doing and saying the same things again and again, of machine-like iteration of automatic responses and semi-automatic actions; days in which human life had seemed to be a swarm of moments whose 10,000 purposes, pointing in different directions, summed to purposelessness such that fulfilment was a kind of 'emptyment'.

As a younger man, he had sometimes blamed others – as doubtless those others did him – for dragging him down into the marshlands of *ennui*. He had cursed all those who had wasted his time in conversations, or rather monologues, that took no note of what he was or what might objectively speaking be regarded as of interest. In such conversations, he was weighed down by the sense that 'we are all any old "anyones",' gabbing, chattering, blathering, and rabbiting, with a fluency born of ignorance and lack of reflection. Ditto time humans wasted in belittling, mocking, even bullying others; in endless point-scoring and tiny victories in little power games.

The most scandalous waste had been in those debased conversations his younger self had sometimes had with himself. For example (to take one random example out of a short list of a thousand candidates) he had once spent a full two minutes feeling cross with a stranger who said 'Excuse me' rather than 'Excuse me, please'. His anger burned on, fuelled by a view of that individual as self-importantly dismissing RT himself as merely an object in the way. Then there was a muttering to himself on a perfect summer evening about something minor someone had said; an unworthy quarter of an hour chasing up malicious gossip about a great musician whom he admired; reading journalism with the intellectual freight of cooking instructions instead of great fiction, poetry, or philosophy; doing something idly – and therefore badly – when he could have done it well so that it would have amounted to something worthwhile (leaving aside the Big Picture insignificance); or criticizing the use of the phrase 'station stop' in train announcements, when this harmless pleonasm was epidemic.

Reminding himself that his life was but a sliver of light between eternity-thick walls of darkness had often proved impotent against the engulfing power of the shallows, and the capacity of yawns to suck out meaning, or to mix a mist of sleep with his thoughts. Tiredness had had its triumphs because drowsiness had seemed a fitting end to each day. Years had colluded with evening to give permission to fatigue, to the view that humankind itself was irreversibly time-soiled, being so many centuries into the tangled, million-stranded conversations its multitudes had had with their many, too-many selves. This had seemed to justify the narrowed horizons, a small sampling of the world that occluded its greatness. The spaces opened up by the birch leaves near his window, the cough in the cathedral, the time-muffled voices from history, snow in Siberia, the lighthouses in the darkness, the far tyrannies and arcane ideas and outrages buried centuries deep – these, he had known, deserved better than a myopic gaze, an inattentive ear. Likewise the immensities triangulated by sideroblasts, stars, and syllogisms. For most of his life, he had been (simply) unequal to his own world, to the giga-crannied universe and the infinitely folded insides and outsides of his days, to the numberless fathomed and visited places of his life and the unfathomed and unvisited places that had outnumbered them and the yet more numerous of the fathomless and unvisitable. How easily had tiredness, approving of itself, made mere pollen dust of stars, and out of dailiness smothered those presences who brought singularities of sense, leaving them like unread books, discarded and crated in the rain.

No longer. Which was why the idea of a right time to die had not impressed RT, though he had had few fears of the far side of death. Yes, some in his world may have been sick of RT but his own appetite for more of himself was generally undiminished until, that is, his body had turned against him and filled his days with malaise, nausea, pain and the challenges of debility. Besides, he felt that he had not had a long life because there was no such thing.

Life is measured out, at one level, in attention spans, or runs of experience, or brief stories. And while the meanings of any given day draw on thousands of previous days (riding on tens of thousands of prior years of human consciousness) the moments do not add up in simple way to experienced (whole) days, or the days to (whole) months. No one can be said to have lived a seventy- or eighty-year span, though the gravestone may report that they did. The continuity of the hyphen linking the birthday and deathday and the co-presence of all its parts is misleading. RT had not at any time in his life occupied his objective longevity. Even to live one day (an entire day) at a time, as those who would wish us to be wise counsel us, is a challenge. The moments of our lives are timeless in the sense of being lived more or less beneath the trellis of the calendar. Or even the clock. The experience of time and the time of experience don't quite know what to make of each other. The moment of pure experience, not bleeding into the future or beholden to the past, not subordinated to a task that makes sense only with respect to a hierarchy of tasks, would have been a mere tingle. Between the tick and the tock there is the escapement through which experience and time make their joint escape.

RT had often been exercised by the thought of *realizing* the length of his life – other than through weariness or sense of repetition. While he could glance back to earlier time-slices of himself and his world, this had rarely yielded the desired result because the 'here' of his present and the 'there' of a remote past were as different as subject and object, and the distance crossed between them like that between one who glances and another who is glanced at. What is more, to touch on an earlier theme, the intermediate locations were not merely receding but absent. When RT had looked at something a long way away, its distance was revealed not merely by the fact that its appearance was small but also by the presence of an intervening visual field. There was no comparable temporal field explicitly intervening between the present and the distant past. Which was why he had sometimes tried to make the distances explicit by populating them with up intermediate landmarks. Unfortunately, this had boiled down to a purely calendrical exercise and did not generate a sense of distances.

The idea of a *long* life was, he had sometimes felt, rooted in a false spatial idea of time – one of the many reasons why he could not make sense of how it came about that a night with a crying baby or worrying over a patient could seem longer than the interval between one Christmas and the next or between his seventh and his tenth year. He had been particularly preoccupied by those experiments with time called 'holidays'. Whereas Day 14 of his fortnight in the Greek sunshine or the Cornish rain had seemed remote from Day 1, the interval between the week before the holiday and the week after was rather less, as if the interval away from home had burst like a bubble under the pressure of resumed routines either side of the leisure-filled sunlit or rain-swept gap. And repetition brought its own compressions: the sunlight of his many thousand days had not distilled to a honey of either time or experience, but was recaptured at best as lighthouse beams momentarily flashing in the darkness that had once been his hours. The class of events to which similar experiences added up represented them all poorly; unlike fallen leaves cumulating to a rich humus.

There had been another reason why he had not been impressed by the idea of the right time to die. Long lives are long measured only against the average: there are longer lives but no long ones. Compared with the eternity of non-existence, even a nonagenarian's death is close to being a stillbirth. All graves are early. Death, therefore, is always an interruption, a lifetime of busyness leaving behind unfinished business. His limits, though expressing metaphysical necessities written into the very marrow of his existence, had been accidental and random in their particular manifestations, and his end all the more unredeemed for being ragged. Even so, he had started to draw some double lines, to sign off this, and wind up that and in a rather desultory fashion to put his affairs in order, in anticipation of the end. In some respects, this was the continuation of the habit of a lifetime of 'tidying up' and also a way of dodging the regret for all that had not been experienced, all that had been undone, un-, or at least under-, lived.

seventeen

Tidying Up

Imposing order on his habitats – filing papers, clearing his desk, classifying his slides, sorting case notes, rearranging data-sets, placing the CD collection in alphabetical order or at least putting the discs back in their cases, labelling this and that – had seemed almost as essential to a tolerable life as laundering his body and grooming his clothes. He had done less than his rightful share of polishing, wiping, and mopping up – not to speak of painting, staining, and sanding. Even so, polishing shoes, putting creases in or out, plumping the unplumped, smoothing clothes or sheets, effacing his imprint from the bed, picking up bits and other ontological proles, consigning waste paper to a basket, brushing off specks, removing gunk, grime and grunge, binning leftovers, collecting peelings, shavings, and sweepings, flossing fluff that had become detached from rugs, jumpers, and other stuff prone to moulting, brushing out corners, sopping up spillage with the assistance of the silent supping of million-capillaried tissues, sweeping paths, calling in aid a melancholy vacuum cleaner to inhale orts lurking in the carpet pile – these had accounted for much of his ado.

While he had found this generally irksome, he sometimes relished correcting disorder and marshalling clutter into an approximation of neatness, often carried out to the background of the radio speaking of catastrophes. Small, achievable perfections afforded substitute satisfaction when real achievements had eluded him or boxes had been left unticked. He was sometimes mildly disturbed by the haste with which on successive Christmas mornings he had cleared up the wrappings torn off from presents which had taken so long to choose, to shop for, to bring home, to enclose in carefully selected paper, to tag and message. It was almost indecent – as if the point of the festival were to get through it as quickly as possible so that the pleasure of restoring the usual order could be savoured.

RT's fits of tidiness had been connected with his profound sense of circling chaos. A tidy house had seemed like a ninety-storey card castle erected in a wind bent on unsorting that which had been sorted. A new stain was always on the lookout for a clean white shirt, the replacement for the broken cup would also break, dust would rain steadily on the just dusted, there would always be hanging threads, lamp bulbs unreplaced, doors that learned to squeak with locks that jammed, and ends forever loose. He was not, however, one of those who would, if offered a posthumous return to the world, have used it as an opportunity to wipe some bird-lime off a windscreen, to deploy a toilet brush, or to polish the silver, tarnished in his absence, or clean out a drawer. Even so, he had found something touchingly quixotic in humanity's expenditure of so much time and effort in resisting or reversing the universe's tendency to become more untidy, in maintaining the less probable order against the encroachments of the more probable disorder. He knew that his *Lebensraum* was not exempted from the mathematics according to which the world had many more ways of being disordered than ordered.

Ultimately, the endeavour to maintain the less probable was

defeated by the limited probability of the emergence and continuing existence of the individual, himself, who had driven it. He who had dusted – what an extraordinary verb – would himself be dust; the face off which he had washed the dirt would become faceless dirt; and that in virtue of which he had been able to remove stains would sooner or later have less structure than, and as much sentience as, a stain.

Much of RTs pursuit of little perfections had been unpleasurable as well as unnecessary. He could no longer think of the hours he had spent fretting over the exact wording of trivial emails and even the choice of greetings in a card, handwritten not once but three times. But seen from the standpoint of the ex-man in the bed, this was as absurd as the worry that a man facing the guillotine might have over a possible punctuation error in his farewell note. But he had also been aware of the temptation of being one of those (dangerous) people who were careless of everything that did not seem to have absolute, or at least immediate, meaning. Sufficient of him, there-fore, was committed to leaving a neat ship behind him; cashing up for an endless, darkless night; at any rate tidying before the ultimate, irreversible untidying.

He had devoted little time to redacting the record, weeding out events that had once had been charged with meaning, so that they did not detonate after his death, like an unexploded device from a war long since over, blowing up a child playing in a field. He knew that any attempt to Photoshop his after-image in the minds of others, distracted by other more compelling things, would have only limited success. He could not influence the discussion – if there were any – about the Life & Character of RT. His reputation would be in others' hands and even the prudent are emboldened to speak ill, or at least carelessly, of the dead.

He would have liked, of course, to have said 'a proper goodbye'. But to whom and to what and when? How far would the catchment area of *adieux* reach into the circles, ellipses, threads of loved ones, friends, acquaintances with whom he had shared the light of day and the dark of night? And as for things, would his farewells include wine, spring evenings pregnant with the possibility of a nightingale's song, lovemaking, likemaking, hatemaking, clouds, reports of Antarctic expeditions, moonlight geometrised by window frames, turn-ups (for the book and on trousers), turnips, car-parking charges, home-work...?

The combination of his sense of impending dispossession and a desire to control the world to the very end and beyond was evident in his wills – his living will (which, signed and sealed, had instructed the world when to leave him alone and let him go) and the will that had instructed the living as to the disposal of his estate. He had for a while wanted to make amends, to apologize to those whom he had hurt, even to confess. He had no God to confess to. Confessing to God is easiest not only because forgiveness is guaranteed but also because he won't be hurt or damaged or further damaged by the confession. An atheist's confession is another matter entirely.

PART THREE | After

eighteen
Last Rites

The interval between his last breath and his funeral is, for those who have lost him, a period in limbo. The clock of bereavement is set to zero only when the last rites are over. For the present, he is dead but the inedible deadstock that once was his, once was him, is mostly intact: those are his arms, these his legs, and there his bald head. No miracles are expected of course; but there is still something of him to be viewed, though no longer anyone to be visited. Conversations and the exchange of impressions are at an end, and the air has transmitted the last pun of his pun-striped discourses.

In this meantime, 'the arrangements' are a merciful distraction from metaphysical reality, covering the granite face of his absence with a foliage of the sort of practical details with which he would have been familiar and, perhaps, a little impatient. It might have been helpful if an envelope marked 'Funeral Directions' had been left in a prominent place. The instructions would have been the first major posthumous exercise of his will, though it would have been dependent on others to be fulfilled, his agency now being entirely outsourced, since his body shares the helplessness of the material world through which the laws of matter operate undeflected by intentions.

It is a strange and admirable piety to respect the wishes of one who can no longer wish for anything. But RT has not drawn up plans for a ceremony that has the twin functions of disposing of his body and expressing the most formal and public of farewells from the world left behind. This may have been modesty. Planning a funeral service might have seemed rather self-regarding, even self-aggrandizing, at odds with the self-deprecating style of the self that has now popped. And it may have seemed like an imposition on others. More likely, it was not the kind of thing he would want to think about on a particular sunny morning or a particular rainy day, especially as he would not be present to experience, even less to enjoy, it. 'Giving *him* a good send-off' is magic thinking because he has already left and is out of range of reports of any ceremony. Better, instead, to imagine a virtual event, at which he would still be represented by this body, this 'it' no longer owned, lived or enjoyed by anyone.

Doubtless there will be some sort of cortege, drawn through incurious streets past pedestrians to whom this *memento mori* will provide negligible instruction, and not much distraction from the littleness and enormity of everyday life. Not all his mortal parts will be inside the formal teak and brass-handled box, the traditional livery in which he is required to attend his last engagement. Bits will have been donated, in accordance with his wishes, in the hope that there will be others able to make better use of them than he. He had liked to think of his body as a potential *smorgasbord* of benefactions, so that those organs with unused mileage would service other lives. If all was well, they would obtrude as little on the recipients' biographies as they had done on his, except insofar as they had made it possible for him to have a biography at all.

The remainder will require burial or, spitefully denying rats and insects their bit of luck, cremation.

The case for cremation had seemed to him stronger. The earth is overpopulated with the dead asking to be visited in body as well as mind. And ashes seem a more complete return to the elemental beginning, cleaner and more direct, avoiding the intermediate stage of carrion, and unspeakable journeys through the digestive systems of ever lower organisms, en route to a final dissipation. His thinking head had not relished the notion of worms dragging their bodies through its orbital fissure or beetles making its tongue wag in something remote from the English language that had exercised it for so long.

No, spare him the insects, spare him anticipation of the horror of a nest of earwigs, a panic of woodlice as a stone is unturned. Or of maggots that had made a cheese heave like a sheet stretched out in a breeze. Spare him entomological encounters closer even than those that sickened him in childhood, from which later life had, relatively speaking, sealed him off so that flies had more often occurred in jokes about Germans' supposed lack of sense of humour than in the soup in which the jokes had located them. As an adult, he had looked with indifference at small creatures madly and ineffectually paddling for their lives and had not felt, as his younger self had felt, the ghastly fascination of minute animate existences bent on being themselves, using their mindless mandibles to further their mindless purposes. So, no to the earth and no to being portioned out between rats, no to being unpeeled to the bone by insects, and yes to the flames.

But first the send-off, the twenty-one-gun salute, or at least water-pistol salute, the farewell in which formal, carefully crafted obsequies will be mixed with more direct expression of non-propositional emotions – with tears trying to find a resting place for their sadness, to express and suppress the hole that is within, and mingled – who knows – with quiet satisfaction, relief or even joy, in the breasts of some.

So, enter centre-stage, the mute protagonist of the ceremony of farewell inside the geometrical box – black as the night that is death's synecdoche, the un-light that stands for an Un the size of a world – borne between mourners irregularly ranked in half-filled pews left and right, bringing them to their feet, asserting his posthumous being as a dead weight on the shoulders of pall-bearers.

Speeches, flowers, music.

Who is to give the speeches – receiving the honour, carrying the burden – requires careful thought to avoid his being the occasion for posthumous offence. Family, friends, colleagues of course; and possibly a grandee, to give the occasion *cachet,* delivering an *éloge* with practised eloquence that betrays nothing of his thought, when the invitation came, that this was yet another addition to the to-do list.(The recently dead do tend to gatecrash diaries). Some speakers will share the nervousness of a best man giving his first public speech on an occasion that has to be risen to. A balance has to be struck between personal reminiscence and formal eulogy, humour and solemnity, the public (or office) figure and The Man I Knew, sadness and celebration, summary and detail. The various aspects of his life, represented by different ladies, gentleman, and children in black, will – they will say afterwards – surprise all present who knew only part of him, saw him in only one of his roles, guises, modes of being.

Those to whom the speeches matter most will find them least

endurable. They summarize, they allude, they praise – but the words remain on the threshold of the occasion, the outer layers of his absence. Even those personal reminiscences, the telling details that show what manner of man he was, will be chosen for outside consumption, to make familiar sense as the kind of things that are remembered befitting the reason we are gathered here today. He cannot, as on so many previous occasions, respond – on behalf of the dead. So the speeches, percussing the boundless silence he has entered from which not the slightest murmur can emerge, elicit no echo. The frail skeins of sound cannot connect the vast spaces he has left behind with the spacelessness he has entered. RT's mourners are talking to themselves.

In the interstices of the formality, sorrow will bring inexpressible memories and a gradual awakening to a new world; a world of loss to be learned, a universe of obstinate absence. Thus the speeches.

'No flowers, please, only donations to a charity' – one perhaps that would support research into the illness from which he died, the untiring efforts of medical researchers to close the portal through which had been expelled from his life. Anyway, his bouquet of choice would have been impossible to assemble, or would have placed an intolerable logistic burden on his once nearest and dearest. For he would have chosen his garlands from the special times and places of his life. No actual flowers, then; only the idea of them, tied into a nosegay of commemorative images. A catalogue of time-pressed blooms, perhaps, to be read out to his mourners.

Goldenrod, that towered above the head of his five-year-old self, when he ran after the others down the little paths they had made on the wasteland next to his home. Syringa to remember the den, constructed in the perfumed hollows of the lovely deciduous shrub in which he passed many childhood hours. Daffodils, pure trumpets of solarity, yellow as a blackbird's bill, awakening throughout his life anticipation of summers to come. Primroses, the colour of sunbathing evening clouds connecting a solitary teenaged walk home from the City library where happiness, born of a sense of possibility, could not contain itself. Purple freesia, whose fragrance filled the hours when he and the yet-to-be Mrs RT spent their first whole day together, walking in Richmond Park. Vetchlings – as yellow as yellow gets – gathered from the sheep-sheared turf of a Cornish headland overlooking the infinitely varied waters of the Atlantic Ocean and the Camel Estuary. And valerian, whose spears had dominated the summer lanes, to be mixed with the lady's bed-straw that had champagned on long June evenings over the tops of the hedgerows. Lilac from a bough seen through the playroom window rocking in the sun-fused April rain as he tried to get his younger child to sleep one strangely numinous afternoon. Corn-flowers because of course they were the colour of her eyes, and on account of the vases filled with them she had left on his study desk. Bougainvillea to speak on behalf of little back streets and baking squares and jasmine ('nightingale of perfumes') to represent the solid warm evenings of the Greek villages they loved together, sea-soned perhaps with a scent of ripe figs, and the music of the cicadas and a fountain teasing the hot darkness with the sound of cool-ness. And, well, yes, roses, if only because his favourite poet had chosen them for his own epitaph, for he had found them to be 'a pure contradiction' since they were 'no man's sleep under so many eyelids'. Such times, such space, time and space itself, gathered in a bouquet of memories, a flood of mingled lights.

Speech, flowers, and, finally – music.

A pause, first, for ironical reflection. Music, he had always maintained, was the supreme art. Unburdened with any responsibility to inform, to describe, even to refer, it could unfold in accordance with laws that were its own entirely, a pure delight, a rejoicing in structures that were built to house a heaven out of the contents of the freest and most untethered of the senses – hearing, which, after all, had constructed the million worlds, actual and possible, gathered up in speech. Being about nothing other than itself, music was at liberty to be about anything or everything. Consequently, it could express and so induce those expansive states of mind that opened up an entire world, the cognate of emotions. Emotions, yes, but emotions perfected.

And now we come to the irony. By virtue of the rich and lovely and subtle links between its beginnings, its middles, and its ends, its progressions and expansions, its harmonies and dissonances, music had seemed to do something to time itself, and make it more human time. More than myths, art for him had been the source of 'machines for the suppression of time'; and music was supreme among these machines. And yet, the contest between time, timelessness, being out of time, and eternity, had always ended with a return to quotidian time – the time that had propelled him hurrying from his messy birth to his messy end. Every experience of music had been enclosed, swallowed up, in ordinary time that seemed at once to be elusive and inescapable, an iron fist that evaporated on inspection to the edge of a wraith of smoke. Music's timeless moments had to be fitted into a schedule, by permissions won from the timetable. Booking ahead for a particular concert had obliged him to be on time and he had fretted at being late. There remained only moments of joy while time continued its work of frogmarching him to his end. And so he would have been on the side of those who, arriving late at the crematorium, annoyed with the traffic that had held them up, find the organist already well into *In paradisum*.

So no music, perhaps. Which might have spared the true mourners – who bear the exit wound by which he departed, as opposed to those who had turned up out of curiosity, duty, or the wish to be present at a mildly titillating occasion – an unbearable sadness at the thought of harmonies which could no longer be shared. Only the *idea* of music, random illustrations from the many miracles of craftsmanship and emotion that had given him delight.

The programme would begin with the perfect dawn captured in the Overture to *Lohengrin*, with the day, fresh and scented brought from a long way off, some place where the entire world was new-minted. Then the Lullaby from The Dolly Suite, connecting his toddlerhood with the toddlerhood of his children, and the old age of his father whose restricted tastes in classical music he had been pleased to cater to. The Gloria from Beethoven's *Missa solemnis* would commemorate those broken nights and early awakenings many decades before when his children, long grown up, had been small and wanted to play at the godless hour of 5 a.m. Next, a carol, beginning as a perfumed mist of sound in cool cathedral air, gathering the light of the dark days of Advent to a candle flame. A song by Mikis Theodorakis that had always seemed elegiac, sad and beautiful, like the sunset refracted rosé through the crest of evening waves on the poignant last day of a Greek holiday. Palestrina's *Missa Papae Marcelli* – a sigh breathed outwards from the Basilica di Santo Maggiore in sixteenth-century Rome into many corners of his life, connecting his study, concert halls, a jet-lagged early morning awakening in America, a multitude of times and places. And, ending a random selection with the appearance of non-randomness, the final *lied* in Richard Strauss's *Four Last Songs*. A quiet autumn evening as the birds settle in their nests and the wanderers think, 'How weary we are of wandering/Is this perhaps death?' Death after he heard his last song, his last sonata, his last quartet, his last symphony. 'He would have liked this' an eternal counterfactual. Music – scented with all the news a summer evening can bring of itself – would have been a mockery: butterflies over a flowery, meadow remote from his darkness.

There it is then: only speeches, redeemed by their formality, the idea of flowers, unchosen music, and, perhaps, a single candle, the soft isosceles of its flame silently fidgeting in the imperceptible breeze, standing for all the light, all the lights, that had filled his days and enchanted him with the revealed beauty of the world.

The funeral moves towards its unimaginable climax: the committal; the disposal of the material remains of him whose death is mourned, whose life has been celebrated. To leave the body unburied or uncremated, to dispose of with contempt – as when the Consul in *Under the Volcano* is thrown into ravine 'like a dead dog' – is a profanity that diminishes us all. To discharge this last respect, humankind spares no efforts. Bodies are repatriated at great risk and expense from war zones so that the Soldier shall not be forever a corner of a foreign field; the ends of the earth are scoured to locate whatever remains of the decaying mortal parts. And yet the dead are disposed of just as if they were 'it' rather than 'I' or 'you', or 'he' or 'she'.

The curtain parts and the coffin, crested with flowers that will share his immolation, advances, bearing its contents – those hands that the eyes knew so well, the feet which he had inserted into daily socks and shoes, the once-smiling mouth – towards the flames, its journey coordinated with the municipal music, and private reflection in the different privacies of those who attend his end. Reducing him to ashes, broken beyond repair, smithereens that could never by smithied to reunion, empty of agency, the plaything of the elements, the remains of one who winked, turned a key in a door, struggled to lift a holiday suitcase.

Speeches; flowers; music; cremation. 'We are gathered here today' – to mourn a death and celebrate a life. 'We' – a collection of people who had in common only that in very different ways they knew or knew of RT – have tried with variable success to rise to this metaphysical occasion. Some have had to resist the temptation to consult watches that have no place in the vicinity of death's unminuted permanence, or sneaked a look the latest electronic device linking this place with the rest of the world. Some will be distracted as this is one of many inescapable commitments in a congested diary. Others will be blinded, by simple grief, to the terrible reality of what has been enacted. The solemnity of the occasion is subverted in so many different ways. Even tears expressive of deep and sincere sorrow make the nose drip and make-up run. This was the motive perhaps behind his one instruction: drinks should have been served at intervals during the service.

There is an air of relief that the cortege, its burden delivered, is now headed for the funeral feast, the reception for the bridegroom of darkness. Ashed, he is spared the knowledge of how quickly the business of the world is resumed. He had sometimes imagined an 'anti-mourner' who has come rocking up to his funeral, chewing gum, and snorting at the more rhetorical parts of the speeches, a reincarnation of the cynical (though much more polite) teenager he had been himself. The idea of the Stoned Guest was not as disturbing as of the reluctant attendees who had wondered whether after all they might skip the funeral; who were already texting in the crematorium garden; and who had rushed off as soon as decently possible. It is as if the crack prised open in life to let him out of the world gracefully had started to close.

But the unscattering prompted by his ending is not over yet. People who haven't met for a long time catch up and compliment each other on their looks or exclaim at the extent to which the children have grown up. The said children are happily playing or quarrelling in the open, the outside, restored to sunshine and the freedom to make a noise and run about. There are awkward reconciliations between family members and friends who have fallen out. The unreconciled observe each from a distance, after minimal greetings, out of respect for the occasion, for the sake of the deceased and those who truly mourn his passing. Strangers discover common acquaintances and common interests. Potential partners are noted, and eyed up: gazes move more slowly than the routine scanning would demand. Business cards are exchanged. The merry-go-round, after this pause, is in motion again. Voices get louder and louder in a kind of arms race to be heard, as the liberal supply of drinks unclenches some and unbuttons others. Before long, the encounter with the end of time is forgotten as tabled time reasserts its authority. Watches are consulted less furtively and people find that it is later than they thought, though not as late as it truly is. The gathering is starting to scatter. The world is growing over the gap he has left, beginning with the people and places that were for him merely penumbral, remote. They now return the compliment by allowing him to disappear over the horizon of their lives. He has had his special day.

The hole which his departure has created is a complex hypersurface, an absence scattered over public and private spaces, concentrated mostly in his house, but spread over unanswered emails and unpaid bills, a mobile phone ringing in the pocket of an empty jacket, an unfulfilled expectation here, a shock there, an 'I didn't even know he was ill', and unfinished business.

Beyond the voices of the departing mourners there is a silence: the silence of the one beyond the buzz and warmth, of a small pyramid of ashes with no point of view. His journey to the Great Amnesia has begun. Rain and wind will do their work and the plate will rust. The inscription will be filled in with moss. The world has seen the last of him. If he walked into a room now, and spoke to us, it would be a miracle. It was no less a miracle last year, but one so commonplace that its true nature was hidden even from himself.

He is embarked on his solitary journey to whatever afterlife, if any, awaits him.

nineteen
Afterlife

Is that it then?

There are reasons for thinking that it is. But there are also reasons for thinking that it might not be. The most compelling should be the least persuasive; namely the impossibility of thinking about your non-existence, if only because doing so presupposes that you exist. RT had been able to think of the non-existence of others – even of those closest to him – but he is not others. That he is the he who is thinking is the fundamental premise of his thought; the springboard from which he thinks about anything at all.

It is also difficult to imagine a boundary, however impenetrable, that does not have something beyond it, that doesn't have two sides facing into different territories. Like many children, he had experienced this impossibility very young when, before he went to sleep, he had tried to think of the end of space, of a surrounding wall that did not face outwards as well as inwards. And the idea of an end that was not also a beginning, or at least the possibility of a beginning, of an exit that was not also an entrance elsewhere, had also been beyond the reach of his thought.

And so when he had peered into death's black mirror, it had seemed as if there were, after all, something to see, or at least to be said, even though what he saw, what he was inclined to say, was borrowed from the world built up, and latterly crumbling, in his living mind. He forms the faint image of an afterlife; more precisely of two afterlives – one (since talk of 'sides' is irresistible) 'this side' and one 'the other side' of his grave.

This Side

Of several things he had been certain. First, the world would not die with him; RT would be survived by X, Y, and Z; others would live on in his absence. Secondly, this being the case, *he* would in some sense continue in traces left in places he had once occupied, and in minds that had known him or of him. These traces would be somewhat randomly scattered, often turning up in surprising locations, and take many forms, most of which he would not have anticipated, many of which he would not have approved. And, thirdly, they would ultimately be effaced, as the bells rung by his name fell silent and a blanket of amnesia cancelled even his absence. Notwithstanding the happy coincidence of his initials, RT would not be retweeted forever.

The first assumption is worth a moment's thought. RT's certainty that the world would outlast him owed its strength to the fact that any challenge to it must undermine the very basis of discussion. If The World went pop when he did, then it must have been a coinage of his mind, and everyone in it, ditto. A world that continued after his death, what is more, was implicit in the idea of his corpse outlasting him, being stored where it may be viewed and grieved over, and being cremated. The belief in his corpse, and the ashes it would become, was an even more powerful counter to the idea that the world was internal to his mind than a barked shin, a rock resisting kicking, or the decline in his bodily fortunes foretold by physics.

Even so, it was not easy to think of The World carrying on (and not greatly diminished, what's more) without his being here, there, or anywhere, and bent on business that concerned him not at all. It was unthinkable to imagine himself absent everywhere and present nowhere, so that sunrise would not cause his body to cast a shadow, 12 o'clock would find him neither indoors nor out, and the evening gather into night without his seeking some way of alleviating the darkness.

Anticipation of his universal absence had been softened by the idea of his being present by proxy in whatever he had left 'behind'; in the wake of his passage through the world.

The most enduring part of his evaporating contrail was his 'estate' to be disposed of, if after all he had not been the first to go, had not pre-deceased Mrs RT; remains that ranged from 'worldly goods' (as if there were any other kind), explicitly bequeathed, to minor effects that did not warrant specific mention in his will; from the house and garden, and the minute portfolio of stocks and shares, and the contents of current and deposit accounts, and of safes and jewel boxes, to worn-out and workaday items, worthless even to a charity shop; and well-used furniture (including the seat that still bore the impress of his buttocks – such are the modes of posthumous presence).

Forgotten things: a wardrobe stuffed with clothes which had once warmed and been warmed by his busy body, now in single file, at attention, literally hanging about, awaiting the call that would not come; books he had read or intended to and papers he had written on or pored over or overlooked; a once-favourite cup. Thus the discoveries of those charged with emptying his house in advance of the estate agent's sign, working their way, over successive weekends, from room to room, up into the loft, and down to the garden shed. A rusty saw that he had last used three years earlier to trim the Christmas tree, a tie that he worn for a joke, a book heavily underlined and spattered with instructions to NB something: to note *particularly* well, notwithstanding competing notabilia. A letter informing him that he had been awarded a place at university, a school report telling his parents that he would do better if he tried harder, and other items bearing news that had once mattered so much because they had opened the path that his life had taken, had been survivors of an earlier clear-out from his mother's house. And a photograph capturing a smile innocent of the posthumous thoughts it would provoke in the person holding the camera. Material to be divided between the children, the rubbish dump and that intermediate zone, the charity shop.

The inventory of his little heap of precious rubbish was endless. The entries had a poignant heterogeneity. It was enough to think of the flotsam he had pocketed in the many pockets he had had over the years. No curator would be equal to the singularity of his hoard. Books pulped, notebooks in the rain, the photographs fading unsorted ('Is that granddad – doesn't he look funny – can we go out to play now?'), and, on a larger scale of obliteration, the garden landscaped and replanted, the house knocked down, the city rebuilt.

So much for the estate. And the wider, less tangible legacy? 'I want to make a difference,' he had sometimes said to impress himself and others, in particular committees interviewing him for jobs. What was this difference? What had been his contribution to the changes that took place between his birth and his death? How had he shaped or at least deflected the course of events? There were the habits, values, beliefs, and ethos of those he had influenced – his children, Mrs RT, his colleagues, his mentees – either in imitation of, or in reaction against, his example. There was the happiness and there was the unhappiness caused by what he did, said, and brought about. There was the keeping things ticking along, or the reforms or revolutions, running the show, or introducing new services, new ways of working, new facilities, identifying new needs. Many of these changes had been reversed or superseded before he had died: the world had moved on and something else was needed from anything he might have had to offer. Yesterday's revolution was today's routine practice and tomorrow's out-of-date ways of working.

To measure a legacy is as difficult as to trace the fate of all those ripples left by stones dropped in a pool; to monitor the trembling of the reeds set in motion by a shower of rain. And the difficulty is greater because the medium through which human waves pass is so heterogeneous, encompassing the material world, social practices, and changed private lives; the joy or misery of children, of a partner, of friends, acquaintances, adversaries, and any number of strangers 'out there'. And as the ripples move outwards so the ascription of parentage becomes ever more speculative. RT as a set of effects will be effaced in the effects of other causes and those that are traced will be spoken in sentences voiced by throats unknown to him.

And he will appear in places he had not anticipated. A daft remark he made once is reported by the addressee to a friend. The friend's family picks it up and it becomes a catchphrase. Years later, two of the children, now middle-aged, quote it to each other, as a way of sharing a memory of their childhood. This tiny ripple, like a soliton, keeps going, boosted at intervals by someone's need to reminisce, to connect with the past, perhaps to affirm sibling solidarity.

If therefore RT looked to the world's memory for an afterlife, he would have also to acknowledge his ignorance of how he would be remembered. He should have expected the unexpected on the basis of the way in which he himself had recalled those who had gone before him. Moments remembering Mrs X by imitating the tone of voice in which she teased her husband, itself an imitation of some general manner of being. Or David Y, whose warm, genial, at-ease manner was one that had influenced him throughout his life. Imitating this manner had lubricated his dealings with others in the unscripted realm of chance encounters. The motif that stood for his father-in-law: someone saying (of a colleague in his office) that 'Old so-and-so was a good sort' in a pullovery tone of voice, with a special kind of tired light in his southern accent, as he had drawn on his cigarette. Would he have recognized this posthumous image of himself? There is no doubt, therefore, that RT would have been surprised – and probably not very pleased – at how he would be recalled or evoked and for what purposes. The distorted afterlife of the famous – whose posthumous existence should have been less the plaything of chance – was a further warning of oblivion to come. Consider the image of the dead poet R. S. Thomas being used to endorse a prize advertising crisps, his 'fleeting look of contempt' directed at the alternative to the prize. On a more cheerful note, Hector Berlioz, who had died thinking that his magnificent music might have been no good, would have been astonished at his posthumous apotheosis, had his body not been beyond even slime. Richard III could not have anticipated the centuries of execration raining down on his name after he had become the eponymous hero of Shakespeare's play; nor that, in the aftermath of the First World War, a Richard III society had been established to clear his name and that they had been active in determining his place of burial after his bones had been accidentally dug up in a council car park and had welcomed the DNA testing that had confirmed the remains as his. Shakespeare? Council? Car park? DNA? The examples were sufficient to instil RT with dread about the way he, a person of considerable obscurity, might be talked about – not because the talk might be malicious or libellous but simply because it would have been point-missing.

His confidence in how he would be remembered was undermined even more profoundly because he had not known, from moment to moment, how he had been experienced by others when alive. He could not even imagine with confidence the reverberations – if any – he had left behind after a dinner party with a few acquaintances. And this uncertainty was more anguishing because he now had no other being than through the recollections of others. How then could he accurately anticipate his posthumous memory? Or anticipate how the memory of his life might influence the lives of those who outlived him? Perhaps those who remained would be prompted by recollecting RT simply to score posthumous points by improving on, clarifying, or just continuing an argument they might have had with him about the role of science in society or the proper way to behave at a dinner party with the Smiths or the just interpretation of someone or other's motives or the meaning of an odd look someone, now forgotten, had had on their face.

Chastened by knowledge of how little we remember each other, we try to protect the posthumous life against the casual infidelity of ordinary careless consciousness. The curriculum vitae is given an additional run in the already-brooded-upon obituary laden with adjectives and anecdotes that direct a course between eulogy and denunciation; and in a memorial service in which he would look in vain for the What It Is Like to Be RT. When a journalist sent Richard Feynman an advance copy of the obituary he had written, he thanked the writer but said he had decided 'it is not a very good idea for a man to read it ahead of time. It takes the element of surprise out of it.' The element of surprise is the source of dismay. How can I be so misremembered?

How do you remember a person? RT asked himself once. How should you? How do you in fact? The phenomenology of commemoration was not encouraging. The thought of his being scattered within and across his scatter-minded outlivers, and the unity by which his moments, days, and years had held together being shredded – was not encouraging. Alive, RT had had a oneness underwritten by, though not identical with, the spatial continuity of the body joining his cranium and his toes and its temporal continuity over days, weeks, years, and decades. Merely remembered, RT's fate was to be a diminishing population of moments lodging in the consciousnesses of others, unknown to one another, a bric-a-brac of words, images, and imagined conversations, retorts even, of splinter-thin angles on him, memories as unrepresentative as the accidentally preserved box-camera photograph taken when he was eighteen months old in which he figured as something between a mite and a misprint. Memories jogged by random association will be jogged by other associations to remember other things – to buy more printer ink, ring one of the living, collect some data – and he will be forgotten: re-burial after a brief exhumation. Centreless, dispersed between a thousand craft that briefly pick up minute fragments of him before returning him to the ocean of oblivion: thus his posthumous fate in the minds of those who outlive him. With the chief and ultimate guarantor of his continuity and his continuing importance – himself – gone, that importance will fade except as an ever-weakening instrument to serve the importance of others.

No wonder we outsource the memory of those whom it is considered proper to commemorate, to memorials with texts inscribed on paper or chiselled on stone, to etchings on trophies, named prizes and lectures, the appellations of streets, or replicate their dissipated flesh in sightless statues, usually forelocked with the cloacal treasure of an indifferent avian – memorials which corrupt, though more slowly and less malodorously than the corpses for which they stand proxy in the park where children are quarrelling over nothing, or by the entrance to museums where the same children have eyes only for the souvenir shop. But these courtesies will secure only a short stay of extinction. The relentless indifference of the centuries effaces the sharpest, deepest letters incised on the most obdurate stone. You are dead a long time, as RT's mother (who had been dead a short time compared with, say, the interval between the architect with his plans and the archaeologist rummaging with exquisite, informed care among the exhumed ruins, or the millions of years unpacked by the geologist from folds in rocks, or of the billion-year-old messages bearing the oldest news of when the earth was new runed by astronomers) had often said. (Though 'the long time' is in the keeping of successive waves of the living. For RT, there would be no difference between his being dead two weeks and being dead two million years.) Burial mounds, sepulchres, cemeteries, the ossuary and the necropolis, introduce only the slightest of pauses into as they, too, wither only slightly less slowly than the flowers with which they were once strewn.

In memoriam will give way to *In amnesiam* and 'RT' will be a lost signal in the noise of the universe. Only a short interval will separate the first phase of his posthumous life – when the Get Well Soon cards keep coming to the hospital, and letters arrive at his address with reminders of overdue subscriptions – and the final phase when heads shake and shoulders shrug in 'Search me' mode as his name is mentioned. Less than a blink of the eternal eye separates the days when the book of condolence is opened, when the last signature collected, when the volume is archived, when it is pulped, and when the pulp dissolves in the universal solvent of change. The last bubble from the wake marking his passage through the world will pop, the last current will lose its identity, the last ripple will break on the shore, into wavelets too small to cause even a paper boat to bob.

Death outlasts all that would preserve the dead.

And yet, 'Lest we forget', out of respect for our common humanity, we say, 'Their names liveth for evermore,' 'At the going down of the sun, we shall remember them.' We stop what we are doing, and stand still and try to obey the command to remember. But more have fallen – casualties of peace and of war – than the unfallen can accommodate. The dead are in competition with the living and with other dead for the wayward attention of those who are busy living rather than remembering, shaping their own futures rather than preserving others' pasts. And the forgetful living will in their turn be forgotten, elbowed aside by the coming horde, as in pursuit of sustenance, pleasure, aggrandisement, and attention, they too sweep aside their predecessors.

This second and final death of the dead had been impressed on RT when, at an annual medical meeting, he had peered in semi-darkness at a slide of those who had fallen since the last meeting and could not make out the names. One day, he too would be an item on an illegible list.

The Other Side

Posthumous existence courtesy of the memory of those who out-live him is not, therefore, very attractive. Hence his resistance to the idea that there is no life for RT beyond the Mondays, Tuesdays, and Wednesdays that he had known. He had been equally resistant, however, to the versions of an afterlife 'the other side' of the grave offered by various creeds. As we look at what remains of RT – the inert, unbreathing, insentient It, as incurious as a stone – it is difficult to imagine a sense in which His I is continuing, discarnate, in some elsewhere remote from this spot, remote from remoteness, outside the space and time in which his corpse, and we who look at and con-template it, are located.

The possibility of some sort of afterlife on the far side of the grave owed what little power it had had over his imagination to his aware-ness of the profound difference between the It in the bed and the He who had lived as I, the distance between the organism that had been RT's body and the person RT who had lived RT's life. One of his most enduring preoccupations had been a mighty gap in our under-standing; namely that we have no idea how consciousness, mind, self-consciousness, the sense of the past and of the future, could have arisen out of, fitted into, and acted upon the physical world to which his body had belonged. A scientist, he had early accepted that science would not be able to offer any explanation of this, not the least because science was itself a late product of conscious, self-conscious, temporally deep agents like himself. Physical laws could not explain how one bit of the material world had formed the con-cept of 'matter' and uttered the word 'world'. And it had seemed to him that mankind could not be entirely a creature of thermo-dynamics if it had been able to conceive the notion of 'entropy'. And so he had been close at times to succumbing to the temptation to think that, if the embodied subject that was RT could continue as a subjectless body, as a corpse, it might also be continuing as a bodiless subject in another place: that his death had been merely the parting of the ways of RT-the-body and RT-the-person.

If he had resisted this temptation it was on the grounds of the asymmetry of the relationship between the two RTs. A body in good working order had been a necessary condition for RT's existence as a person; but RT's continuation as a person had not been a necessary condition for the body to be in some kind of working order. After all, his body could have continued beating, breathing, growing, if RT had done time in a personless coma. He had never been a mere lodger in his body. His tenancy was too close to identity.

Even so, while he was inextricably part of the physical world, he was also apart from it. We have unpacked some of the great spaces between RT the embodied subject and the complex biological entity courtesy of which he was able to live a human life. The laws of physics that brought the organism RT into and out of being have a deeply puzzling relationship to the person RT lived, narrated, imagined, and ultimately lost. So, perhaps we are not entirely justified in seeing the subject of the embodied subject as simply the manifestation of a certain kind of organism (*H. sapiens*) in a state of health. We have reasons, perhaps, to entertain the idea that that our possibilities are different from those available to pebbles, trees, or even chimps.

We should not under-estimate our collective ignorance. RT, *in utero* in 1946 had had no idea of the world he was to enter and strut about in, so knowingly, for a while. It is not impossible that this world has itself been another womb whose walls successfully muffle the rumours of another kind of reality, one perhaps that is even wider and even brighter. That he had shared this second womb – the world he has left – with millions, indeed billions of talking, thinking, remembering, investigating, discovering, fellow humans, may not fundamentally alter the fact that he may have entirely misunderstood his own nature. After all, secular thinkers, including RT himself, were entirely confident in their belief that previous, pre-scientific ages had got many fundamental matters wrong. And it is not unreasonable to conclude from the history of human consciousness that all ages of mankind are equidistant from any complete truth about human nature and, in particular, how humans fit into extra-human reality. According to the very knowledge that makes it difficult to believe or even imagine that RT can in some sense outlive his living body, humanity is an infinitesimally small part of the creation which it is trying to understand. What is more, this kind of knowledge itself does not seem to be part of the material world that made RT's existence as an organism possible.

It is possible, therefore, that RT was not, simply and without remainder, identical with, and inseparable from, the mere organism whose self-envisaged death has prompted this meditation. Much that he had lived for, the meanings that guided him, had been bigger than his material being. Even though that material being was the condition of his having access to those meanings, he accepted the possibility of a different kind of light, illuminating a new landscape of thought. Perhaps the last of the darkness that came from his particularity would wobble like a black flame and go out, leaving the field to unchallenged light. At the very least, we may permit ourselves to imagine RT journeying to another mode of existence, as if his evident end had been the gateway to a new beginning. He himself would of course have been sceptical, as he had been of all claims to knowledge of death's 'dataless' night.

The Pythagorean idea of a 'transmigration' of his soul seems perhaps too empty – except insofar as it is filled out with angelic images, like be-gowned Chagalian shooting stars jetting through the night sky – and the term 'soul' is too laden with problematic connotations and contested meanings. There is another image that allows RT the indulgence of feeling sorry for his posthumous self and, indeed, for the selves of those he loves: the idea of a Journey to the Island of the Dead. It is a continuation of that Journey to the Sunless Land in which his world shrank and the future closed down, every fastness was unfastened, every lock unpicked, all defences stormed, and he had seemed to be getting ever further from the places and times where and when the world was home and he was at home in the world.

It is an embarkation, as if extinction would be followed by transition and transformation. We may imagine, entirely illegitimately, a clammy overcast day as he is rowed to the Other Place. There is a fat swell, muffled oars, and solemn silence. We think of a faint afterglow of RT moving through relative calm after the alarums and excursion of the Intensive Care Unit, of the flashing lights and beeping machines, and after the pain and the panting and vomiting and fitting has ceased.

It is difficult not to think of him as afraid – though shorn of a body to cower or a heart to beat fast – lonely, and disorientated, stripped of his seniority and know-how and *savoir-faire*, without office, role, career, wisdom, acquired sophistication, bravery and prudence (all equally beside the point), heading for a destination beyond the use of tears, to a place where the cry for help dies away in infinite silence, a state to be endured with all the unknowing uncertainty of a baby, helpless, lost and afraid. A baby thrown into intimations of awful adult solemnity; a metaphysical rookie anticipating a process of initiation he cannot conceive, his condition equally inconceivable to those – above or below – who are grieving his departure, imagining they are thinking about him, when they are, of course, facing away from the RT who is no longer.

The entire world he knew has involuntarily turned its back on him and, thus cast adrift, he can feel in the very marrow of his being the incommensurateness between his self and the great universe. Infant-naked, he has no wallet, keys, credit cards, clothes, authority, standing. He is absolutely unequipped, lacking even skin or arms or eyes or mouth. How much he will remember of himself in such circumstances and how much his expectations will be shaped by those memories it is impossible to say.

The model of a journey, crossing not merely time zones but passing out of time to something other than time, an eternity that is neither before nor after time, and arrival at a destination that is neither above nor below, is a way of unpacking, making graspable, the idea of the transition from a world we know in part to another of which we know nothing, the elusive notion of continuation in being despite utter transformation. It is, of course, unthinkable. And so is the notion of any life lived in a space empty of material objects (including his own body), of heres and theres as he has hitherto understood them; neither light nor dark, sounding or silent, fragrant or without odour, touchable or out of reach. A life independent of, shorn of, the thousand intertwined narratives that populated his days; without hours or clocks to record them. A life directed to no conceivable ends or purposes, without anything corresponding to the knowledge that guided and misguided his life. A sense of a self which, notwithstanding disembodiment, is somehow replete, not hollowed by anticipations or memories, hopes or regrets.

We have nothing in ourselves which would enable us to host the idea of such a life. Though we may mobilize the words that correspond to it, it remains a place of words, of concepts beyond conceiving. Any content the afterlife may have, therefore, is borrowed from what is seen in a rear-view mirror that populates its surface with images and ideas drawn from the past.

And this will be one of the many ways in which the afterlife will be a knot of contradictions, the most striking of which will come from portraying an eternity of repleteness using material drawn from a time-torn limited life of insecure and incomplete meanings. Another will be reconciling the loss of individuality – no egos, conflict, or competition – with continuation of the RT forged in a seethingly polycentric world of embodied subjects helping, hindering, or just bumping up against other embodied subjects.

RT's language and imagination grew out of shared life, voices speaking in a common air to common purposes. It is scarcely surprising, then, that there is no idea of an essentially solitary existence in death that could be imagined or articulated. Yes, there are words that seem to straddle the two sides of the grave; and there is one – 'God' – that has the draft to accommodate both death and life, promising speech outside language. But, ultimately, such words are spoken by humans, and serve human purposes that do not reach outside our life. That may have been why his strongest intuitions as he approached death were streaked with terror, as he imagined the confusion and helplessness of the disembodied new arrival in a place more alien than the strangest, least welcoming reception class. Fear stronger than the hope of a gradually completed awakening, organized as a new kind of day is organized, not fragmented into mere epiphanies and episodes, brighter than anything RT had hitherto known and exceeded by the same degree as his ante-mortem consciousness had exceeded that of an ant.

There is nothing to be relied on in speculations as to the nature of any afterlife RT might enjoy, or endure, after he has disappeared from sight. On what principles could he draw in building up an image of a next life, given that, notwithstanding the universal patina of knowingness, we know and understand so little of this one? We may imagine the nightmare possibility of entering the judgement of One whose gaze would synthesize the immediate private knowledge RT had of himself, and how he was experienced by all those others whom he had encountered in his life, with a view from nowhere, neither parochial or partial or even situational, that did him objective justice, and Who would see the overall balance of his kindness and cruelty, of care and carelessness, of conscientiousness and idleness, of honesty and cheating, competence and blunder, of rectitude and transgression, of decency and nastiness, totting up the net good or evil he had contributed to the world, and presenting him with a true picture of the difference he had made to it. Could he bear the self-knowledge that would come with that gaze?

RT had sometimes imagined a philosophical version of Paradise. He is in a brilliant sunlit dawn by the sea. He washes his face in a spring and then, radiant with the simple and profound joy of conscious existence, he take his place at a table in a little cell, and listens to his thoughts as they swell to an ever-widening praise of the mystery of the world.

But forever – and alone?

Coda

While we cannot waken the dead, the dead may awaken us.

The Black Mirror is not my debut as a corpse. At the age of thirteen I made my one and only stage appearance, as a slave boy, in a play whose name I have forgotten. I had been 'slain' (offstage) by a despot. My onstage appearance was entirely passive: I was to be carried in by the hero, a doomed rebel (who in daily life was a house prefect, authorized to wear a gown, and bear much delegated authority) and be publicly mourned. My loinclothed body was mahogany-stained to give me an authentically Arabic hue. An impassioned speech – 'What has thou done to this innocent child?' – was spoken by Prefect Jameson over my lifeless frame, as I lay on the boards, endeavouring to make my breathing invisible even to the parents in the front row.

The speech brought the first half of the play to an end and won enthusiastic applause. When I opened my eyes, I discovered that I was in front of, not behind, the curtains and still visible to the audience. There was scattered tittering. I raised my dead body and walked off the stage to a second round of applause, echoed in a subsequent review in the school magazine in which I was congratulated on being the most lifelike corpse the reviewer had ever seen.

This time, the corpse I have played is my own, in an endeavour to look at life – my life, your life, anyone's life – from a virtual viewpoint outside it. I hope, reader, you will understand why there has been so much about RT and about life in the foregoing pages. He is a mere exemplar, picked out neither for any particular merits or failings. The autobiographical is anthropological: he is offered as an instance of *H. sapiens. Mutato nomine et de te fabula narratur.* 'Change only the names and this story is about you.' (Horace, *Eclogues.*) The stuff of his life is the stuff of yours. He, like you, has spent a portion of his life tying his shoelaces or getting cross at being kept waiting or worrying about a child with a fever or terrified of death or enjoying lying out in the sun.

Using the idea of the Nothing beyond his life's end to illuminate aspects of the Everything that precedes it, *The Black Mirror* has been an invitation to marvel at all those seemingly important hurries, all that activity and passivity, action and experience, from the standpoint of a stillness in which all hurry is spent, time is no longer tabled, insentience rules, and all ado is adone.

I have over-delivered on my promise to give an entirely inadequate account of Life in these pages. While the headings bear witness to an attempt to impose an idea of order, there remains something of The-Philosopher-As-Bag-Lady in the hoovering up of this, and that, and the other, into a single text. The endeavour to decant Life on to these pages is not dissimilar to ladling an ocean with a shallow teaspoon. Normally, salience governs selection; but the standpoint of death subverts the hierarchy of the salient and the irrelevant, the important and the trivial. There is no defensible end to the number of themes that might have been addressed. And each theme – patches of light, the manifestations of ownership, the non-Euclidean geometry of suburban spaces with their multitudinous partitions, the transformation of breathing into speech, and the togetherness and apartness of the world and our lives – could be written about at much greater length.

My fragmentary obituary has left out much of the greatest importance in the world. The wordy fever called RT was lucky enough to have lived out of reach of the great historical events of the century in which he had passed most of his life, and large-scale tragedies were for the most part mediated through print and screen. But this is not the only reason why the focus has been on the quotidian, with little or nothing on violence, madness, war, injustice and the fight for justice, on imprisonment, enslavement, and destitution, even though for many these are daily, even birth-to-death, realities. Consuming sexual passion, gratuitous cruelty, the various ways in which we lose our freedom, are hardly touched on. There have been no arguments for justice, no denunciations of brutality, no calls to arms against physical, psychological, social, and political despotism. There is insufficient about the most important events in many human lives: love and marriage; and having children. My defence – apart from a respect for my own privacy and a desire to avoid subjecting readers to trial by photograph album, iPad, and reminiscence – is that my aim has been to remember and reflect on the basic elements, the aspects of living daylight, that extraordinary events rest on; to seek our depths in places where they have been concealed through being so intricately folded over; to unpack them from the miraculous coherence of a smile with a shake of the head in a person walking with necessarily interrupted purpose to post a letter or answer the phone. The fragments I have gathered are, I hope, sufficiently disparate to triangulate the great spaces in which those things not touched on in these pages become possible, and find their occasion and sometimes terrible reality. Our ordinary hours are richer, deeper, and larger than we conceive or imagine them when we are *in media res*. It is in the interstices of the 'long littleness' of life that the treasures of non-destitute existence are to be recovered and the great events – ferocious conflicts, sexual passion, and moral crises – have their origins.

So while *The Black Mirror* is an attempt to reveal some aspects of the inexpressibly rich and strange nature of human life, it is the product of a gaze darting hither and yon, alighting here and there. RT's butterflying visits to his world do little justice to the complexity of the decades, the years, the days, the instants of his, or anyone's, life. Anyone who has kept a diary knows that it is impossible to bring the world to the page. Journals seem closer to reality the more remote, and hence forgotten, that reality is. A true and comprehensive account of something as simple as an afternoon walk – what was seen, noticed, said, remembered, felt, thought, done – could never be written. As for the afternoon, or the holiday to which it belonged, or his children's childhood to which the holiday belonged – well, the most accurate and honest entry would be, 'Forgot what did (saw, said, felt, thought).' Life writing is engagement in an aerial reconnaissance that at best captures the shape of the fields and the grid of the streets but misses not only the aphid climbing up the stalk of grass but also the meaningful glance between two people passing on a staircase.

This may be adequate for what is, ultimately, a work of praise and gratitude, trying to present the world as seen through the eyes of a wondering love that I feel I should experience less intermittently; a vision that looks past our fate, past the mindless energies of the world, past love and war, to joy. I would like to think that this may inspire someone to battle harder against those things that make the lives of many filled not with wonder but with suffering.

Let us stand once more in front of a mirror, mindful that the item we are looking at will in due course be our corpse. Listen to the clock, amid the sunlit bruit of our busy hours – the phone ringing, the train to catch to the vital meeting, the to-do list growing ever longer – ticking more loudly against the silence of death's Grand Negative. And then, awakened by death, let us bid farewell (or, alas, *au revoir*) to our respective corpses; come back from the dead to change the world or our lives, or simply to relish that it is still possible to put the kettle on, look out of the window, and exchange smiles with another human being.

Le vent se lève!…I'll faut tenter de vivre!